THE *Profit* PROCESS

Building a Successful Business Without Profit Leaks

BRENT TRENHOLM

The Profit Process: Building a Successful Business without Profit Leaks

Paperback ISBN: 9781946824622
Hardcover ISBN: 9781946824639
Ebook ISBN: 9781946824646
Audiobook ISBN: 9781946824653
Library of Congress Number: 2019916578

Published in the United States of America | INDIEGO PUBLISHING
www.indiegopublishing.com

Publisher's Cataloging-in-Publication Data

Names: Trenholm, Brent.

Title: The profit process : building a successful business without profit leaks / Brent Trenholm.

Description: [Deltona, FL] : IndieGo Publishing, 2019. | Includes 5 illustrations. | Summary: Uses research, shared experiences, and case studies to demonstrate how to build a more prosperous and successful business by focusing on the customer while avoiding pitfalls and profit leaks.

Identifiers: LCCN 2019916578 | ISBN 9781946824622 (softcover) | ISBN 9781946824639 (hardcover) | ISBN 9781946824646 (ebook) | ISBN 9781946824653 (audiobook)

Subjects: LCSH: Business planning. | Industrial management. | Profit. | BISAC: BUSINESS & ECONOMICS / Decision-Making & Problem Solving. | BUSINESS & ECONOMICS / Development / Business Development. | BUSINESS & ECONOMICS / Management.

Classification: LCC HD31.2 .T74 2019 (print) | LCC HD31.2 (ebook) | DDC 658.1--dc23

LC record available at https://lccn.loc.gov/2019916578

indiego PUBLISHING
Our Brilliance . Your Success
WWW.INDIEGOPUBLISHING.COM
WWW.INDIEGOPUBLISHING.CO.UK
WWW.GETINDIEGO.COM

The Profit Process: Building a Successful Business without Profit Leaks has received several great testimonials and reviews from renowned leaders in business, industry, and entrepreneurship:

Kevin Harrington, original shark from the hit TV show *Shark Tank* and creator of the infomercial:

"Brent Trenholm will motivate you to take a hard look at your business and start taking action to correct your profit leaks. He's also able to help you find what's not working and guide you to make great changes."

James Malinchak, featured on ABC's hit show *Secret Millionaire*, top-selling author of the book *Millionaire Success Secrets*, Founder of www.BigMoneySpeaker.com:

"Brent is an amazing speaker, author and business leader. I love his book! If you're ready to achieve more business success, then read and absorb the strategies in this brilliant book by my friend Brent Trenholm!"

Brian Tracy, President of Brian Tracy International, author of *Now, Build a Great Business*:

This powerful, practical book shows you how to generate more net profits from your business than you ever thought possible. Get it, read it, practice it—today!

Gerry Foster, Gerry Foster Branding, Home of the Big Brand Formula:

"In this era of hype, simplistic how-to tips, and marketing ideas, Brent Trenholm's book stands out for its depth and wise personal insights for profit optimization. He demonstrates the critical need to understand the importance of identifying and plugging profit leaks that every business has to deal with. Brent's perspectives on the need for every business owner to build or transform an existing business into one without a 'leaky bucket' so to speak should serve as a wake-up call for many owners. His guidance, recommendations, and suggestions are nothing short of exciting. *The Profit Process* is truly breakthrough work!"

David Dean, President of David Dean and Associates:

"Brent Trenholm's book *The Profit Process: Building a Successful Business without Profit Leaks* reveals his mastery in the art of making changes towards running a successful business."

Scott Schilling, executive coach, international trainer/speaker, and best-selling author:

"Brent Trenholm's book *The Profit Process* is loaded with examples of common profit leaks and business challenges that many never identify or fix. If you want to add to your bottom line immediately and make more money, put Brent's genius into action.

Joe Theismann, President of JRT Associates Inc. and World Champion Quarterback:

"I congratulate Brent for being able to pinpoint those nasty business things that always annoy us. I especially enjoy the simple and reasonable approaches he takes to correcting those."

DEDICATION

Many thanks and hugs to my wife, Caroline Létourneau-Trenholm, for her intuition and guidance throughout this process. I couldn't have accomplished this milestone without her.

TABLE OF CONTENTS

HOW'S YOUR BUSINESS DOING?

Entrepreneurs and business owners know that starting and running a business is challenging, yet we enjoy the thrill and sense of accomplishment when we open our doors to the public, whether those are physical doors or virtual.

Not long after opening our doors, however, another set of challenges begin. Every company encounters problems, and when you're the business owner, it feels like you're the only one struggling to stay afloat. Let me assure you, you're not alone.

Certain profit leaks are common to all businesses; two of the most common are the owner or manager's inability to lead and the employees' unwillingness to be led. The distinguishing difference is the level of severity of the problems and what you do to solve them. Are you willing to make the required sacrifices to maintain and/or build a vibrant, collaborative business?

This book will serve as a guidebook and a direction finder based on my forty-plus years in business. I've included many examples of common employee and departmental problems that I've encountered. For each example, I've provided tips to help you decide whether that problem exists in your company and how serious it is to you and your company culture and profit margin. Then, I illustrate why this problem is a profit leak in your business, and I provide solutions and next steps for turning those leaks into profits.

Throughout this book, we will explore the internal aspects of managing a business and creating change. We will discuss common best practices, and we'll take a close look at options to lower your company's operating costs and increase your productivity. If that seems like an impossible task, rest assured, it can be done, and when done correctly, the positive results will surprise you.

As you read this book, reflect on the following:

- Why are certain aspects of my business so successful?
- Why are some aspects of my business struggling and possibly even failing?
- Why am I not able to bring in the profits my business needs?

Every business has profit leaks, but not every business owner is aware of them; or, if they are, they don't know where to look for leaks or how to change company culture to plug them.

Before you read further, I invite you to think about your business and ask yourself the following questions:

- How do I uncover my hidden profit leaks?
- How do I decide if my profit leaks are numbers and dollars or people and systems?
- Which departments or people are costing me money and lowering my profit margins?
- Am I a profit leak?
- As the business owner, do I give my employees enough freedom to do their best?
- Do I share my business vision and convey my company mission, or do I resist because I want to be in control?

With those questions in mind, let's discover how to build a successful business without profit leaks.

CHAPTER 1

Good Business, Bad Business: What's the Difference?

We're all familiar with the various ways that society pegs you as a certain type of person. I was pegged early on as different. When I was young, I did not like school except for shop and PE classes. Everything else seemed like a waste of time. In the years leading up to my twenties, I was a quiet thinker who observed everything, including what people did and how they did it. I eventually gained a better understanding of why certain things happened. Then, when I was in my twenties, I began to realize that people listened to what I had to say. I had always liked building race cars —the faster you can get there the better—and when I did anything, I looked for the shortest, fastest, and most efficient method and route to get from Point A to Point B. It just made sense to me, and it made life easier. I approached my career with the same laser focus on the intricacies of efficiency: what it is and why it's important.

I am an action-oriented doer, as evident in the things I enjoy in my free time: I'm a sports fan, mechanic, leader, coach, and a bit of an athlete and overall active person. When I was younger, whenever a project caught my interest, I launched right in with gusto, just to get it done, regardless of my time and energy or personal sacrifice.

I am very organized and energetic, and I approach problems from a practical and logical perspective. As a strategist, I think like a chess player, several moves and counter-moves ahead, to ensure forward momentum with the fewest mistakes possible. When I work on projects, I determine the result I want to achieve before I get started. That way it's easier to put the various steps in place to accomplish the goal in the least amount of time. I systematize just about everything I do, from cutting the grass to managing processes and companies.

Based on what I have observed and learned from work and life experience, there's a three-step process to getting anything done efficiently:

Step 1: Initiate a conversation and create an ongoing dialogue.

Step 2: Take proactive steps to accomplish your goals by planning a defined set of responsibilities and actions.

Step 3: Create accountability steps and detailed action plans.

I have written this book for both types of business owners, those who are ideas-oriented and those who are action-oriented. Both are needed in any thriving business. Together, they can harness the key aspects of the profit process to avoid profit leaks and maximize profitability.

To build a solid, thriving business with profit as your goal, you need a foundational understanding of the following key concepts:

- Know what your company does.
- Learn the importance of your role in the business other than being the leader.

- Understand how the various processes of your business work, from the initial sale to delivery, and how those processes affect all employees.
- Know the objectives and goals you want to attain, and share them with your employees.
- Be able to adapt and adjust quickly as the business climate shifts and the economy fluctuates.
- Communicate regularly with your team members, customers, clients, and suppliers.
- Master the ability to sustain the flow and sequence of the processes unique to your business to maintain profitability.

IT ALL STARTS AT THE TOP

In any entrepreneurial enterprise or business, everything starts with the decision makers. This is the prominent area where ideas are born, standards are created, and comprehensive plans are developed and implemented. In contrast, the inability to develop and communicate a comprehensive plan and to set realistic standards is where profit leaks begin.

The leader's ability to communicate is critical to the success and upward trajectory of any business. This includes asking questions and responding to them, understanding what you want to achieve, and being able to explain it in detail to your team. Everything starts at the top and flows down, good and bad.

I had a great conversation with a colleague about what this looks like from a structural perspective. Historically, upper management sat at the top and directed traffic without getting too close to the activity. We call this top-down management:

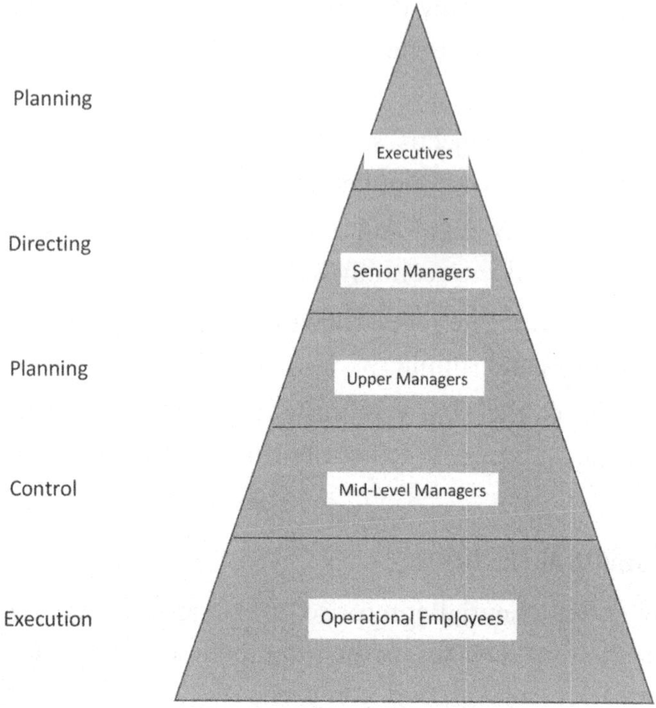

"Top Down Management"

This type of manager is effective in a working environment where s/he makes all the decisions and decides how all aspects of the business will be run. The top-down manager is the guru at the top of the mountain who has all the information and only shares it with those who "need to know." This works well in a "Yes, sir" environment. In a lot of cases, however, this single layer of management could be reduced or eliminated, and the business would be just as effective.

Some managers use a more interactive approach and demonstrate a more effective understanding of their employees' challenges and successes because they work closely with them. This style of management enables everyone to be more engaged, proactive, collaborative, and interactive:

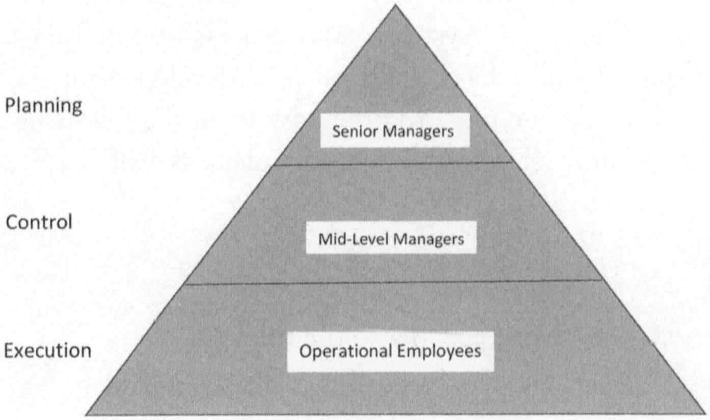

"Flat Management Structure"

The interactive management model allows information to flow back and forth, within and among the organization, in a more efficient manner than top-down management. For some businesses, this efficiency in communication can reduce the need for excess staff, thus cutting overhead and creating greater cohesion in company culture. The communication cycle is more collaborative, actions are put in place, empowerment is promoted, and more people have a vested interest in the business, which in turn creates a positive cultural change and results in fewer profit leaks and greater profits.

BUILD A STRONG TEAM

Another challenge that many business owners or managers encounter is being willing to let go of authority, but this is vital. When you understand the importance of delegation and teamwork, your business will be stronger. As a result, you will progress through the profit process as a natural part of your growth. The seamless cohesion of delegated teamwork builds a safety net that is strong enough to prevent profit leaks.

Let's consider the synchronicity of a baseball game as an example. Baseball has a defined system and a process of play that is unique to the sport. Certain basic skills must be developed for the team to be successful. At the most rudimentary level, the following must be learned and understood before you can play baseball:

- How to throw a ball
- How to catch a ball
- How to play according to the rules
- How to effectively use strategy and teamwork
- How to adjust quickly during game play
- How to communicate effectively with team members, coaches, and umpires
- How to score and win the game

Baseball is all about teamwork. The same is true when managing a business. You are as strong as your team's ability and willingness to participate in the growth and success of your company.

As entrepreneurs, we are often "the company" when we start out. As our business grows, we realize we must hire talented people to help us carry out our vision because we cannot do everything by ourselves. The philosophy we use in choosing our team and the culture we create is indicative of whether we scale the upward trajectory of the profit process, or lose our footing and suffer profit leaks as a result.

What should we look for in the people we hire?

- An appreciation of the diverse skills, talents, and abilities among their coworkers.

- Leadership skills and the ability to communicate effectively.
- Openness to being coached and willingness to improve and change as needed.

KNOW WHAT YOU WANT TO ACCOMPLISH

I've coached countless managers and business owners who admitted they had completely lost focus on what they wanted to achieve. When you don't know what you want to accomplish, you don't know how to attain it. Without a clearly defined mission statement and proactive measures to achieve that mission, you are operating within a broken business model. You wake up one day and realize that your business has fizzled out and is limping its way to the finish line. Worst-case scenario, it probably won't survive unless you decide to take the steps necessary to create and embrace change.

If this sounds familiar, it's time to ask yourself some honest questions:

- How do I work through the daily challenges that arise in my business?
- How willing am I to make sacrifices when the need arises, even though it might upset someone?
- How willing am I to hire talented employees and trust them to do the job I hired them to do?
- Am I so focused on staying in good favor with my employees that I lose sight of my role as their leader?

Whether your business is service-oriented, production-based manufacturing, or professional, you need to be proactive in certain key areas to avoid profit leaks. If you can honestly answer "no" to any of the following questions, this will reveal the gaps in your business where profit leaks can occur:

- Does my company have defined structures, systems, and processes? If so, are they followed?
- Do I have specific written requirements for how my clients are obtained and treated?
- Does my company ask a specific set of questions of each client so that we can better understand their needs?
- With the goal of fulfilling our clients' needs, do we have a protocol in place for how the information is gathered and distributed to our teams?
- Do we follow a proven process, method, or structure to take the transaction through the entire profit process from sale to delivery without repetitive re-work?
- Administratively, do we have the right people in the positions best suited to their strengths?
- Do my employees have the required skills and abilities to do their jobs?

It is essential to have a detailed plan or system in place if you want your business to succeed. If there is no plan, your business will struggle, and most likely, it will eventually fail.

My driving motivation for writing this book is to help you prevent unnecessary failure and experience ongoing success as you navigate the profit process so that you stand out as a leader in your industry.

Take a moment to think about your business. Assess it honestly, including the people in it. Are there any gaps in your processes where profits are leaking steadily, and you don't know what to do about?

Be encouraged! The fact that you are reading this book and thinking about change shows that you care about your business and the people in it. You want to succeed, and you want to know exactly how to maximize profits and minimize leaks. You are ready to navigate the profit process. Welcome aboard!

CHAPTER 2

The Profit Process: It Begins with the Customer

Although the profit process seems straightforward—you price your products and/or services high enough to cover your business expenses and to gain a surplus without pricing yourself out of the market—the reality is that earning profits is a multi-faceted process with several parts and pieces that form a cohesive whole, like an interlocking puzzle. Each person and department within your business is a piece of the puzzle. They interact with other departments, employees, and customers throughout the production and/or service cycle. The key is to manage these interactions so that employees and departments openly participate in the profit-making process but do not manipulate the direction of the business or the overall planning. Allowing others to manipulate you or take control of your business is the surest way to suffer even more profit leaks.

Revenue and profits are by-products of a strong company culture that welcomes and promotes effective systems and structures based on communication and collaboration, with excellent customer service as a primary goal. Everything else is secondary.

Everything starts and ends with the customer. This is true whether your business sells goods or services. When the customer chooses to order or purchase something, the profit process begins. From that point, the customer's order passes through a series of steps. At each of these steps in the process, your customer is the fulcrum on which everything balances. When you get to the last step in the profit process—delivery of the product or service—you will have a satisfied customer who will bring you repeat business.

The profit process is a continuum that involves all the departments within the business, starting with sales and overseen by management. The following graphic accurately conveys the importance of this interconnectedness:

As depicted at the top of this graphic, the profit process begins and ends with the customer and his or her order for a product or service from your business. As each order travels throughout the various departments that make up the typical business, profit leaks can happen.

In the course of your business operations, a thorough exchange of information with the customer and staff, both written and verbal, will ensure the successful delivery of your service or product. Without continual communication, you will lose customers, confuse staff, and suffer additional self-inflicted profit leaks.

Throughout this book, we will examine the power and effectiveness of having systems in place and how to build them to produce profits. Systems are the foundation of every successful enterprise. When your orders are processed using efficient systems, the information is shared with relevant people and departments, and revenue is generated. Every employee and department affects the outcome of the order. If your orders follow a sequential process through the systems you've established, your business will succeed.

Let's examine the concept of structures and systems in more detail.

STRUCTURES AND SYSTEMS

When you brainstorm an idea for a new business venture, or you consider ways to refine your existing business, everything starts with a blank piece of paper and expands from there. On this blank piece of paper, you write down your dreams, and from those, the concepts of your business structure begin to take shape. From those components, the structure and direction of your business begins to emerge.

Establishing the structures and systems of your business involves much decision-making. You will need to outline the defined path your business will take to sell your products or services.

- Who will sell your product or provide your services?
- Who will take care of the administration for what is sold?
- Who will create the components of your products and services?
- Who will manage the components of your daily business operations?

- Who will deliver the service?
- Who will take care of the money coming in, pay the bills, and account for all financial transactions?
- Who will oversee all of the above?

When you've answered these questions, you've established the structure of your business. Within that structure, you will need to create systems to effectively and profitably manage the workflow. If one or more of your systems are out of sync, the structure of your business will collapse.

In most businesses, the structure will consist of several systems. Many businesses are comprised of four basic systems: sales, service, delivery, and administration. Within these systems, there are specific roles, responsibilities, and internal business relationships.

You are at the helm of the business structure. Using a boat as an analogy, you are the rudder. Without a rudder, the boat will go in circles. You cannot steer a rudderless boat or make informed navigational decisions. Without a rudder, you cannot chart a course and steer toward it. You will go nowhere fast.

A system provides the foundation of how you do everything. It can be as simple or as complex as it needs to be, but you need a system. One of the most common mistakes that business owners make is to establish a system that is so rigid and complicated that the workplace culture and productivity are negatively impacted while costs escalate due to redundant or useless work. Your systems should be clearly understood and functional to your employees and to those on the outside observing your business. It should be evident that your systems are working.

Strive to build a practical system for every activity in your business. As you brainstorm ideas, think of the methods, processes, and steps that each activity within the daily operation of your business will follow to sell your products or provide your services. Pare those down to the most vital components, and post them where employees can see them. Build every aspect of your business around the utilization of these systems. Train your staff to follow them to the letter. Help them understand that when they do, the process from sale to delivery will flow smoothly and seamlessly, with few hurdles and interruptions. This is how to build the profit process without suffering leaks.

The structure of a business typically has **three** components:

- the **system**, which defines how the structure works
- the **process,** or the road map of the structure
- the **elements**, or the individual tasks and events that make the structure work efficiently

Some structures have **four** business components:

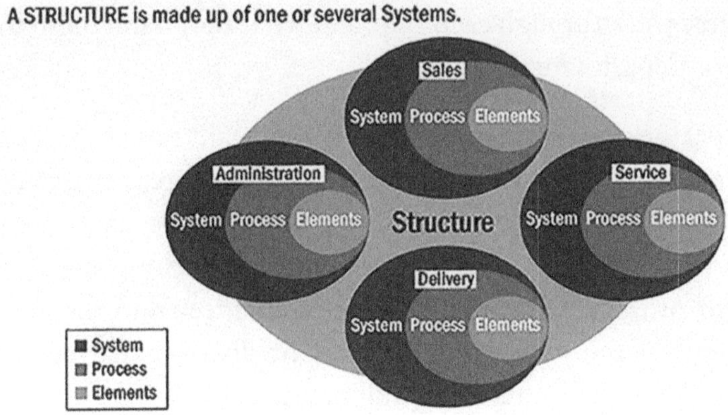

The structure of your business is built on a series of events from sale to delivery. If one of these events is missed or skipped, the system fails and profit leaks happen.

Teach your staff to understand their responsibilities and objectives within the scope of the product(s) and/or services you offer. Provide boundaries and direction to build their confidence. Your staff will soon recognize the fundamental connection between their actions and the exchange of ideas with their coworkers and management.

How do you know if your systems are working properly? The following example maps out a method to review a system.

Consider an order you completed recently. Follow it through each step, each person and department, from inception to completion. Do the same activity with a brand new order: follow its pathway from your first contact with the customer to the final follow-up email, survey, or whatever method you use to know that your customer is satisfied. This strategic routing review should lead you though a direction-finding sequence that enables you to see and experience current pathways while observing what happens during each step and who touches the information, how many times, and why. This will also help you understand why your employees respond a certain way while they work through each step, and what they hand off to the next person or department.

When performing these exercises, it's important not to gloss over the effectiveness of your systems. Diagnose which areas need adjusting (they usually reveal themselves), make an honest assessment, and build a plan to fix them. Examine the steps, pathways, and hurdles that your orders *and* your customers experience from the beginning to the end of the transaction. You might discover that certain vital steps in the process are being skipped because employees consider them trivial, or there is not enough time to get them done. If this is happening, discuss the concern at an appropriate time to help your employees understand why this system is in place. When you and your team discover what needs to be adjusted, you move closer to achieving consistent and exceptional results.

Much of what I do as a consultant revolves around a five-step process:

1. Write down your specific goals and objectives.
2. Assess the current state of your business in its life cycle.
3. Create an action plan and use it as the roadmap of your business.
4. Implement the action plan. By doing so, you will make forward progress.
5. Follow through and follow up. This will involve accountability and client care, because after you make changes, you might need to make some adjustments. Make sure your business is flexible and fluid enough to embrace and implement the changes specific to your industry.

Now that we've discussed what it takes to build a strong company with a solid profit process, let's take a closer look at a case study to demonstrate these concepts.

CASE STUDY – INCOGNITA CORPORATION

A fellow business consultant referred me to a company that she had worked for who was having significant operational trouble. Although she provided some insight as to the problems she had observed, I talked to the owner of Incognita Corporation and heard his rendition of the problem areas in his business and the changes he hoped to bring about. I was hired as a consultant to begin the process of creating positive change within the company. The owner of this medium-sized business was having trouble with the overall flow of processes and procedures, which negatively affected his profit margins, and he lacked understanding concerning the direction he wanted to go. The owner assured me he had all the systems and processes in place, but I soon realized that this was not true.

After I spent about four hours walking through the facility and talking to people, it was evident that this company was suffering profit leaks on multiple fronts. To put it simply, their business model was broken.

Because everyone did a little bit of everything, the employees often interfered with each other rather than helping each other. When I asked one employee, "What are you responsible for?" he responded with, "I get sh*t done." This was emblematic of the general culture at this company.

Many people and departments at Incognita Corporation had a habit of working within their own walled garden and did not collaborate with others share important information on a consistent basis. Their specific aspect of the business was all that mattered to them. There were limited formalized processes, and those were not followed. When new, streamlined processes were created, there were comments such as, "We already have that process," yet no one knew where it was documented, and the complainers were usually the ones who did not follow the process. The result: this company had created a twenty-year culture of bad habits and selfish acts.

Poor communication, especially among the management team, led to lack of guidance and direction on the shop floor and among the management team. What was said was often completely different from the actions taken. This simple act of double talk fostered mistrust among staff.

Some employees received special treatment and were rewarded for no reason, even when they did not meet the objectives of their position, their department, or the company. This was especially true for employees who were favored. It is a fairly common practice in most businesses for management to treat employees differently, but it sets a negative precedent. If favoritism occurs too frequently, it erodes a company's culture.

Overhead was too high due to a mismanagement, wasteful spending, and lack of planning. The parts manager often forgot to order a vital part in a large order, and he did not follow up to ensure orders arrived on time. Receivables were not checked, stored, or accurately recorded to ensure parts matched what was ordered, and there was no inventory control.

As a professional consultant, it was my responsibility to present the facts, provide options, and take direction on what the client was willing to do. In this case a structured purchasing procedure and inventory control was created and shared with the responsible employees, including assigned tasks and accountability steps to ensure parts were purchased on time and expedited. Furthermore, every incoming part was identified and cross checked and assigned a recorded location and put there.

Extreme egos and abusive behaviors were allowed to run rampant, which created a negative, complacent atmosphere with "told you so" comments fired back and forth when everything fell apart and everyone was scrambling to fix problems at the last minute. Everyone assumed that someone else would take care of a particular task, and they blamed others when problems resulted because those tasks weren't done.

Certain members of the sales team made special deals with clients, but they did not share pertinent information with the rest of the team. Then, they insulted their coworkers when the job was not done properly, or it was late. Mediocre quality and poor accountability standards seemed the norm at Incognita Corporation. To resolve this, accountability standards and professional protocols were implemented and enforced.

There were three different cultural segments within the facility, and each had its own perspective on deliverables. All of them considered their ideals as top priority. One segment thought everyone else was incompetent, another constantly complained about most things yet offered no solutions, and a third culture segment just wanted to be left alone so they could work. Trust issues were rampant as many departments blamed the other departments when problems occurred. There were fractions within fractions.

The owner wanted to improve his company's profitability, but he was hesitant to make the required sacrifices, or to have someone else do the hard work involved to bring about change. He allowed a few chosen staff members to dictate what they wanted to change, even though he knew where most of the bottlenecks occurred. It is a very dangerous profit leak when a business owner hands the reins of their company to the most vocal, radical contingent among their staff, resulting in organizational chaos. This chaotic approach had become such an intrinsic part of the twenty-year culture of Incognita Corporation that it was considered a normal part of operations.

Most staff knew that something had to give. There was a lot of support for making changes but a few key people were playing both sides, meaning as long as change and/or accountability did not affect them, they were fine. After a few discussions, I realized that these "certain key people" refused to be coached nor did they want to be part of change.

The result of this walled garden mentality: nine silos operated within the facility. The profit leaks that resulted varied from one or more of the following problem areas.

The administrative staff went from order to order and task to task, fixing mistake after mistake, putting out fires instead of implementing proactive systems and following them. There was a general lack of computer literacy in the office. Parts and salespeople wrote everything out by hand and had others type the information into a computer system, as opposed to simply using the templates provided or the ones in the system and creating the document at the beginning of the job order. The results of these actions caused data entry errors, which resulted in finger pointing. This was a huge profit leak. A detailed process was created and shared with a directive to follow the process, but unfortunately, the same two problem employees did not comply. The owner continued to complain yet blocked every attempt to help the two problem employees improve their work habits.

This normalized reactive culture had been entrenched for years, which meant that creating change was a difficult task, but with willingness and commitment, it could be achieved. As a professional business consultant, I made the following recommendations to Incognita Corporation:

- Develop a personal job description for each employee, or ask employees to provide a written description of their job, then discuss. Your agenda is to have employees know their duties and not interfere with others.
- Formalize processes and procedures by taking inventory of workstations and the ways in which information and tasks flow from one to the next.
- Examine the office/building design. If it does not facilitate the easy sequencing of events, consider ways to restructure it for optimal workflow.
- Put in place a plan to make the building more conducive to serving the industry that your company provides.
- Ask an expert to come in and do a working session on how to share important information with coworkers.
- Develop a dependable process of communication from management to employees. Document this process and follow through with it.

- When it was time to discuss management issues, I explained that although it is not uncommon for people to be treated differently in a workplace, favoritism sets a precedent and erodes the company's culture, not to mention breeds trust issues. I left it at that. There is an entire chapter on creating a healthy workplace culture later in this book.
- Incognita Corporation's operating costs were too high for this type of business. Facility costs needed to be reviewed with senior managers to see where wasteful or redundant activities were taking place. I led them through this exercise, and when profit leaks were exposed, I led a strategy meeting with senior staff to brainstorm solutions. Another aspect we explored was to find an alternative source of revenue, such as selling parts or renting out office space.
- A faulty parts department and poor inventory control consistently created costly overruns. Take stock of your inventory and the amount of inventory you carry at all times. Year-end write-downs force you to make painstaking and costly decisions. A comprehensive purchasing and inventory control system needs to be refined, implemented, and followed.
- The two problem employees (Parts Manager and Sales Lead) needed to be supported in personal growth and provided with two options:
    - Be open to process change, and be willing to change their practices accordingly.
    - Be willing to let go of problem employees to make room for skilled team players.
- Office staff needs to be supported in upgrading their computer skills. Handwritten documents need to be phased out over a two-month period. No exceptions.

These recommendations were necessary for building or rebuilding the company's culture while improving profitability. This is true for any company in any line of business. Management and employees need to exhibit three qualities:

- Eagerness to learn
- Willingness to perform
- Willingness to be held accountable

The most effective way to build or sustain any successful business or organization is to examine what is being communicated, not only among employees but between the company and the customer. We will take a closer look at communication in the next chapter.

CHAPTER 3

Communication: Everything or Nothing

At every stage of the profit process, communication breakdown is the most common downfall. Ineffective or lack of communication is the number one problem in business today, and will most likely be number one in the future. That does not mean we should not make an effort to rise above the norm. Despite the tools available to us for instant communication via phone calls, emails, texts, social media updates, blogs, and website updates, people are flawed human beings, and we all make mistakes. Each of us has a perspective that seems correct and right. It can be challenging to step away from our vantage point and see things from another's perspective. This should not mean that we ignore problems when they arise, however. It means we should address each problem on its merits and create a solution. Your job as the manager or business owner is to set a positive example of collaborative, open communication within your company's culture.

At its heart, the purpose of communication is to understand one another's perspective. You, your staff, and the customer all want clarity and understanding about transaction expectations, whether it is the sale and delivery of shoes, designing a website, running a doctor's office, or running a manufacturing facility.

When we look beyond the basics of management activities and business operations, and focus our attention on the delivery of our product or service, we notice that clear and open communication keeps everything flowing. Good communication starts with management, and the order cycle starts with your sales department, but it doesn't end there. Selling a product or service is one skill. Sharing the pertinent information to the people who create the product or provide the service is another. To best serve your customer and ensure that orders are processed quickly, efficiently, and accurately, your salespeople must be able to communicate the client's exact specifications and expectations, and to do so verbally and in writing.

Lack of clear and open communication is the number one cause of profit leaks. If this problem is not dealt with early on, and if there are no systems in place for how and when your employees will communicate with each other or the customers at every step in the profit process, you will fail.

Running a business is challenging. It gets tougher when people and/or departments want to be exclusive entities, often referred to as silos in the business world. A common problem in any business is employees who do not follow directions or standard procedures, and do not regularly share pertinent information with their coworkers. Often this is the result of personal beliefs and/or abilities, past practices and experiences, or not knowing what to share or with whom to share it. Without a formal structure, continual learning, and functional systems, a business is bound for failure. This is a management profit leak.

Since the profit process begins with the Sales Department where excellent communication is so vital, let's look at an example of the type of salesperson who overpromises and under-delivers, and who overlooks the value of communicating with various departments at each step in the profit process.

CASE STUDY - JOE'S MACHINE SHOP

Joe's Machine Shop needs Pete's Manufacturing to produce some Blowout Preventers for them. Joe has a meeting with Ken, the lead salesman at Pete's Manufacturing. Salesman Ken feels confident that Pete's Manufacturing can get these products manufactured at a cost to the client of $450,000 with a delivery date of six weeks. In his over-confidence, Salesman Ken feels he can save time by not consulting with the rest of the sales team or with the production and delivery departments. Salesperson Ken forges ahead and makes all the arrangements with Joe's Machine Shop to close the deal. He signs a commitment letter and sends the order to Manufacturing. When the order is processed, Salesman Ken realizes that the actual cost to complete the order will be $500,000 with a delivery time of eight weeks rather than the $450,000 cost and six-week delivery he promised. To cover his mistake, Salesperson Ken blames the Manufacturing Department for not being able to make the product using his projected costs and delivery time. The company loses $50,000 on the order and is two weeks late on delivery.

What are the problems in this scenario? Do you see any of these problems occurring in your business?

Most companies have a set structure for sales, which includes getting all the departments together and discussing costs, timelines, and scheduling—in other words, open communication to make sure everyone is on the same track. It seems Salesman Ken didn't feel it was important to follow company protocol. He had limited exchange of critical information with the appropriate people and departments and made the final decision on the order without ample consultation with those who needed to produce and deliver the order. His intention may have been good, but his lack of communication and his decision not to follow protocol made everyone look bad.

Salesman Ken overpromised and under-delivered. This is a huge mistake. For the Sales Department to generate revenue, all members of the team need to meet and finalize a workable plan before accepting an order. This also means the manager sits in on the meeting to ensure that all departments have their say concerning what works for them and their teams. Inventory must connect with vendors to ensure the necessary parts are in the facility when manufacturing begins. Accounting needs to let everyone know the true production cost of the sale. That means they need to know the actual costs of personnel, overhead, resources, and materials to complete the manufacturing of that product. Manufacturing must schedule their work within the parameters of all the other team members' steps in the process. After all of that has been done, the salesperson is free to present to the client a workable estimate of the product and an accurate time frame for completion and delivery.

Let's examine a couple of solutions to help your business deliver what you promise.

All orders need to include detailed information that is relevant to all departments. It's crucial that the sales department have a formal information package that travels with the sales order. Whether your business provides a professional service or is a production facility, it is essential to gather and record the following essential pieces of information about the customer and the sale:

- Customer name, contact person, pertinent e-mail address(es), phone and fax numbers
- Client purchase order and delivery date(s)
- Description of what the client is paying for and what you will provide

- All technical information—drawings, directions, special deals, and any other relevant and pertinent details
- Shipping address, contact person, phone, email, and fax

The entire team must be informed of every milestone in the profit process so that everyone is on the same page. Communication and cooperation are critical to this process. When this information package is not passed along, cost overruns and late deliveries become standard practice, and key people in every department scramble for pertinent information.

In the earlier example, Pete's Manufacturing lost $50,000 and delivered the product two weeks late. This could have been avoided had Salesman Ken made it a priority to communicate with the rest of the team. Lack of clear and consistent communication causes profit leaks and frustration at every level.

The following is an example of an effective communication process during the selling of a product or service. This example can also be used for a variety of company types because the only difference between businesses is what you are offering:

- The salesperson goes to the client armed with thorough product knowledge.
- Communication (both written and verbal) with the potential client takes place.
- A firm understanding of what the client requires is confirmed by the salesperson and is communicated to the manufacturing or service department.

I recommend having a comprehensive document that outlines specific pieces of information to ensure everyone understands what is being ordered. This is essential, and it reduces the chance of profit leaks right from the beginning.

From this communication, you will gain a firm understanding of the costs to create the product or provide the service. These costs include the following:

- Human resources requirements and availability
- Direct costs of materials and tooling
- Overhead: what it costs to run the department or company Workers' Compensation Board, insurance, utilities, assets amortization, royalties, and lease
- Labor: straight time and overtime. If overtime is required, the costs go up.

From the information exchanged, a sales price and its corresponding profit numbers are worked out and an accurate delivery schedule is set.

- Salesperson goes back to the client, presents the offering, negotiates the final particulars of the offering, and receives a purchase order.
- The salesperson creates the customer information package and hands it off to the manufacturing or service department—whichever will do the work.

Many companies say they have this type of protocol in place, yet the need for clarification runs rampant. Every company that I have coached does one of two things: they demonstrate a lack of systems, or they have so many systems that it is challenging for their people to be creative. Every question and delay decreases profit and diminishes staff and customer satisfaction. Take the time to obtain accurate customer information from the beginning, communicate it to each department throughout the profit process, and you will suffer few to no leaks.

If you, as the manager or business owner, don't have a formal information package that travels with the sales order, this is a perfect time to set one up. As we saw in the example above, without an informative hand-off package, cost overruns and late deliveries become standard practice because employees make assumptions due to the lack of information and no time to track it down. This poor practice costs profits and increases frustration. Knowing exactly what you must deliver is of prime importance to achieving success!

Something else you may want to consider if your business is in manufacturing, construction, or repair: information packages smeared with dirt, grime, and grease are hard to read. Plastic-covered iPads with voice recognition software might work better for your field workers. The information can be printed once it's submitted to the appropriate department.

Incorrect information is another common problem that can cause a lot of frustration as an order moves from sale to delivery. Documentation is so important, especially the accurate kind. Poor synchronization between departments has the potential to disrupt the progressive flow of information and documentation. I have consulted with several businesses where people or departments tried to process an order only to discover that pieces of information were missing, and/or the order was rushed through, and steps were missed. Incorrect information usually results in cost overruns of 20 percent or more. If your team does not provide detailed information at each step—what the customer has ordered, delivery date, customer contact person, and supporting documents—you are sunk before you start. The following is a sample from my business that you can download from my website:

PRO-B Management Inc

Calgary, Alberta, Canada
info@pro-bmanagement.com
www.pro-bmanagement.com
403-519-4330

Sales Contact Information Sheet

Salesperson		Phone	
		Email	

Company Name			Order Date
Customer Address		Phone Number	
			Extension
Contact Name		Cell Number	
		E-mail	
Title			

Shipping Address		Purchase Order	
		Delivery Date	
Shipping Contact		Phone Number	
		Email	

Scope of client request: Create attachment or use back of sheet if extra space is required.

Additional Information: Provide a list of supporting documents and attach.

This sheet is to be completed in its entirety to name, label and explain what the client wants. Any agreed-upon variation needs to be documented on this sheet.

Next we will examine some typical and frequent communication pitfalls in every business. The first has to do with meetings. We've all been to those meetings where one person monopolizes the conversation and does not allow others to comment, respond, or ask clarifying questions. It is impossible to strategize and ask or answer questions when everyone is waiting for the monopolizer to stop talking. In every successful business there needs to be a firm understanding of the protocol for meetings where there is an agenda and minutes are taken then reviewed by someone (usually the manager) and shared with everyone within twenty-four hours. If participants don't depart the meeting with assigned tasks and firm timelines and deadlines for completion, you've defeated the purpose of the meeting.

The length of time it takes to complete the exchange of information, get all the thoughts out on the table, and put a plan in place to strategize towards actionable tasks and next steps is a true profit leak. Let's examine the very real cost of disorganized and unproductive meetings.

Suppose a weekly meeting has eight department heads and two support staff present. These meetings usually last about an hour and a half. Let's say that the eight department heads earn an average of $95 per hour, and the support staff earn an average of $50 per hour. This gives us a total cost of $1,290 (wage and benefits) for everyone to attend the meeting. Now, because the meeting went in circles and was inconclusive at the end, productivity will decline by roughly 20 percent for the week for each department because questions were not considered nor answered, tasks were not assigned, and decisions were put off. This company has 100 employees with an average weekly payroll of $48,000. Therefore, 20 percent of this $48,000 is $9,600 of lost productivity. This is a serious profit leak and too often ignored!

The meeting moderator should come to the meeting with a detailed agenda and minutes from the last meeting. All participants should receive a copy. Even better, to save time at the meeting, the moderator should email the minutes from the previous meeting to all participants a day or a few hours ahead of time so they can arrive informed and ready to proceed. A good strategy is to expect the participants to come prepared with some form of actionable item from the previous meeting.

The meeting moderator can use popular meeting structures such as Parliamentary Procedure to structure meetings for efficiency and order. The most important features of any meeting are to discuss ideas and be willing to exchange information. The moderator's task is to keep the meeting on track so that each item on the agenda is addressed and resolved.

Create an atmosphere that encourages participants to listen to each person speak until it is the end of their time to talk. However, the moderator should have the ability and finesse to guide the monopolizer to listen to others, respect their perspectives, and allow them to speak. Often people listen only to pounce in and respond with what's circling through their mind as opposed to taking the time to really hear and understand what is being said.

Paraphrasing is a powerful and effective skill to use in business meetings. Suppose the speaker has just made a point about an item price, delivery time, and projected profit. When it's your turn to speak, you might say, "So, if I hear you correctly, you are saying that if we sell the product for ten dollars with a two-week delivery, we will lose two dollars per item?" The answer will confirm or clarify the information presented.

As the business owner or manager, take an active role in the meeting as a speaker and as a listener. Hear what your team members are saying; help them understand the business perspective from management's vantage point. In doing so, you will allow everyone to get clarity on items, overcome any obstacles, and achieve the company's objectives.

Ask questions to clarify or understand. "Can you elaborate and explain your second point, which I understand to be…?" Clarification ensures that everyone is on the same page and all are moving in the same direction.

Invite feedback to your perspectives and opinions. This should be a direct invitation: "After I share my perspective, I would like to hear your thoughts on possible alternatives and improvements." Even if you have an open, approachable management style, introverted employees might be hesitant to offer their opinion or feedback unless specifically requested to do so.

On another note, it is extremely important that your employees enjoy their work environment. Lively dialogues and friendly laughter are welcome sounds in any workplace. Workers who listen to their coworkers' perspectives demonstrate respect for others. This respect raises the bar for everyone while enhancing company culture.

Setting discussion parameters is good business practice, and it allows the quiet participants a chance to be heard. With a set agenda as well as monitored and effective communication, you can enjoy constructive and effective discussions without suffering the profit leaks of wasted meeting time. Meetings are rewarding when everyone returns to work with an understanding of the goal, accomplishment, or expected result, and they know the roles they will play to bring that to fruition, including assigned tasks and definitive timelines.

Another communication pitfall is the workers or clients who complain about others and have difficulty accepting blame when situations go awry. For example, you may have employees who hang around the water cooler or make frequent trips to the breakroom for coffee to discuss last weekend's sports game. Fifteen minutes go by, then half an hour, and most of the issues have yet to be discussed. The cost of this wasted time is about $1,000 when you calculate the average of each employee's earnings per hour.

Or, during a coffee break, Bob talks about Jill, Jill talks about Sam, Sam complains about Fadia, and half an hour later, all they've done is generate new complaints to bring to the manager, or worse, to "share" with others and create rumors.

Time-wasting unproductive conversations destroy team culture and perpetuate negativity. When the manager comes around and asks Bob how he's progressing on that paperwork, Bob is vague and elusive with his response. "Oh, I can do that off the top of my head. We can sit together when you're ready." Another tactic is to deflect the problem onto someone else. "Oh, I thought that was Jane's task." Be clear about your expectations: "Okay, when you've completed the preliminary outline of that project, drop by my office. Let's plan to get that done before lunch." Sometimes you need to give specific direction: "No, that was your task, as discussed in the last meeting, and I expect you to get it done by three o'clock this afternoon."

A key aspect of the manager's role is to allow everyone to be heard, even the complainers. The manager can help those challenging employees make their point more clearly by offering guidance and direction. The complainer needs to know that the manager will proactively find solutions to any problems that arise. It takes training and willingness on the complainers' part to change their habits and find effective ways to communicate their woes.

The understanding between the complainer and the manager should look like this: Manager maintains a true open-door policy, which encourages the complainer to request a meeting time (and duration) to discuss issues with the manager. The manager informs the complainer to come prepared with concise and professional notes. The complaints must be business oriented, respectful, and must include solutions. The manager sets the tone of the meeting and insists that the talk be about the problem and possible solutions, not about other people.

Keep the discussion focused on the issues, not on personal problems. As soon as we make a remark about someone's personality during a conversation, it gets personal, and people take offense to such comments. When we focus on the issues and get away from you and me, it is much easier to dissect the problem, examine it from a fresh angle, and put it back together better than before, with complete agreement.

The manager should allow a certain amount of time for the conversation, and then shut it down at the end of the time limit or if the discussion goes off track. If the complainer does not come prepared, and only comes to your office to spew negativity, take the time to listen and clarify the main problem. Maybe the person's points are valid and just phrased poorly, or maybe the employee is the wrong person for the job and in the wrong workplace for the wrong reasons. You may have to find a new role for that employee, but sometimes, you must allow them to grow by letting them go.

Here's a common scenario: the senior manager complains about results, yet when his employee asks for clarification about his assignment, the manager does not answer, or he simply says, "Figure it out." If this happens, the manager has no right to complain if the employee takes the initiative, makes a decision, and the manager doesn't like the result. Partial blame should be on the manager for not clarifying expectations or desired results. Surprisingly enough, the blame game happens more often than most are willing to admit because blaming another is easier than looking in the mirror and thinking about what you could have done and how you could have produced a better result.

If you are unsure how to proceed, you have the right to be persistent in getting clarification. This goes for the employee and the manager, but the manager needs to clearly state his or her expectations, not only at the beginning of the assignment, but at every step along the way. When providing direction to my employees, I like to say, "It's your plan and our result. Ask all the questions you want."

Effective communication not only takes place in person but in writing and in phone messages. When someone leaves you a voicemail or sends you an e-mail or text message, they expect a response unless they say otherwise in their message. Not responding to phone calls, emails, or text messages is a faux pas in communication. It's as if you're telling that person, "You're not important to me"—not the message you want to send if you hope to grow your business.

Good communication etiquette also means being open, engaged, and receptive during the entire conversation.

On another note, if your company has an open-door policy, it must be respected. When managers hide in their office behind a closed door, it's a sign that they are in over their head, they're seriously disorganized, or they're indifferent to the overall health of the company. It's not a good habit for anyone to isolate themselves from their coworkers and staff members. The entire team needs to know what's going on. A two-minute conversation between you and your employees keeps everyone informed.

Keep forward momentum going with effective communication. It saves time, money, improves departmental and company culture, and invigorates everyone on the team.

Remember, the communication cycle starts with management and flows from the top down. It spreads laterally from employee to employee and interdepartmentally, and the positive atmosphere impacts the customer or client. I invite you to create a thorough and robust process for information to pass through all the necessary departments from sale to delivery. This is one of the most important ways you can improve profits and minimize leaks.

Coach your staff on how to share important information with others and to be flexible and collaborative. Establish a communication-rich culture and a positive workplace environment. Be a good communicator. Treat your employees and junior managers with the dignity and respect they deserve, because they are key components in the profit process and the company's overall success.

CHAPTER 4

Workplace Culture

If you want to build a team of exceptional performers, create a collaborative environment that supports people from a variety of personal and professional perspectives and backgrounds. This is accomplished in two phases. First, you need to develop a culture based on trust and mutual respect, which in turn creates collaborative performance-based workforces. Secondly, it is essential to maintain commitment and direction on your mission statement.

To create an effective, performance-based workforce, let your employees be their best. Support and encourage that environment. Avoid the common mistake of making arbitrary changes just because you can as the CEO, manager, or business owner. Many companies fail for this reason, even when everything is flowing along at a record pace. This alters the progressive course, productivity halts, and profits are reduced. To use an old saying, if it's not broken, don't fix it. Let your team do their best work without having to constantly adjust their workflow to appease management's latest whim. Micromanagement is a huge profit leak. Let's consider an example.

At Polar Bear Industries, the facility manager made tremendous strides to reduce operating costs and increase profits. A year and half later, the company hired a new CEO who had a very different philosophy on how to run a facility. His first decision was to remove the facility manager. Result: forty people quit over a three-month period, profits disappeared, and the facility became a revolving door for managers and employees. Was the CEO's decision based on ego or logic?

Hire the person who can quickly see opportunities to improve efficiency and who demonstrates the vision and ability to make progressive, unbiased decisions to transform the business. Be wary of the managers who have grandiose plans and want to make big, sweeping changes without the research and data to support their methodology. Quite often, these types of managers come in, cause chaos, and depart, and you are left to clean up the mess.

INTERVIEWING

This brings us to the all-important exercise of interviewing and hiring the right people to create the progressive culture you want in your business. Interviews are a very important (albeit time consuming) exercise in figuring out the applicant's strengths and weaknesses, and whether their habits and beliefs will contribute to your progressive culture. A bad hire could mean the person ends up being let go a couple of weeks later, which is not beneficial to the company or to the person who then has to search for a new job.

As an experienced hiring manager, I have encountered all sorts of job applicants. Some of them led directly to increased profits in my company. Unfortunately, a few others were examples of potential profit leaks had they been hired:

- A young man came in with his buddy and said I should hire them both even though I had only advertised one job vacancy. Had they presented themselves as exemplary workers, I might have considered it. The reality was that neither was a good fit our company's culture.

- The person who looked and smelled like he had not had a shower or washed his clothes for a week. First impressions are very real in the workforce, and this person gave the impression of being so lazy that he didn't even care about his personal appearance and cleanliness.
- A young lady continually exposed her cleavage, and demeanor suggested that she hoped it would work in her favor and land her the job. As the hiring manager, make it clear that you are looking for the right person with the right skills and ability for the job that will fit the culture you're building in your business.
- The person who came in drunk. If this was supposed to be the candidate's best presentation of himself, there was no place for him in any professional workplace.
- The young man who arrived an hour late and made a scene because he expected to be interviewed despite his lack of concern for our time. Imagine what this person's time management skills or work ethic would be like, assuming he could make it to work on time on a regular basis.
- Lastly, the person with an outstanding resume who did not have the skills mentioned in her resume, which we discovered when we gave her our standard preliminary hiring tests. I wondered how much she had paid someone to write that exemplary resume.

Before any job posting goes up, the hiring agent needs to ask him or herself two questions to determine the purpose and scope of the interview as well as the hiring decision. If you're the hiring manager, ask yourself this question: *Am I just filling a position, or am I looking for the right person for this position?*

If you are filling a position, your approach should focus on the resume and its content. This approach is fast and easy. If the person does not meet one of the requirements—having an MBA and a minimum of ten years' experience, for example—the applicant is disqualified. This method does not take into account transferable skills and natural ability. Remember, there are companies and people out there that write incredible resumes. The candidate might look brilliant on paper, but after they're hired, it becomes glaringly obvious that they are not a good fit. This is true whether you're hiring employees, supervisors, or managers. Knowing what you are looking for is paramount.

During the hiring stage, it is vital to determine how they attained a specific academic achievement or job title, and how they gained the work experience described on their resume.

The challenge of this approach—hiring just to fill a position—is you have a greater chance of creating a department full of technically skilled individuals with identical skill sets who are more focused on "I" instead of "We."

If you are hiring a person for the position, your emphasis should be on the applicant rather than the resume—their approach and attitude toward the job, and their ability to uphold high standards—though a well-written professional resume is also important. Your goal is to determine whether they can uphold your standards, and if not, this should be a deal-breaker in terms of offering them a position. As you examine their resume, talk with them during the interview about their likes and dislikes, as this will give insight into who they are as a person. Check their references and examine how they have applied their education and experience to determine their willingness to learn and the types of transferable skills they offer.

An example of this is to consider the various ways that a person with twenty-plus years of experience has evolved in their career. Does this candidate have all the qualities you are looking for? If so, great! But the next question is, can they willingly contribute to the team? Even if they have decades of experience as well as multiple skills and abilities, are they willing to learn the specifics about your business? Do they understand how they must work as a member of the team to achieve your mission statement?

This approach is more time consuming because you are trying to read between the lines of each applicant's responses to build a shortlist of people who will fit into the company culture you are building, and who will contribute to the upward trajectory of your profit and success.

HIRING

Now that you've interviewed a slate of qualified candidates, it's time to hire the best ones for the job and for your company culture. A common mistake among managers, and especially among small business owners, is to look for someone just like them. My advice is rather tongue-in-cheek: If you want workers to give you answers you always agree with, and/or you like to hear your own voice parroted back to you, then yes, hire people like yourself or people with lesser skills than you. But with that I offer a caveat: If this is your usual mode of operation, I guarantee your business will not make the desired gains you wish to see, and you will eventually fail.

Hire people who will make you better than you are, and who will help you drive your business to greater success. How can you determine that when you don't know how the person will perform until they've been hired?

For the best results, it is crucial to know what type of skills and personality traits you are looking for in a new hire. Hiring people based on their skills as opposed to their resume is an excellent start, because those skills will transfer to new learning. You can always teach them the particulars about your business, but they need to demonstrate the foundational skills you require for the position.

Industry research and my experience as a manager have proved that skill diversification creates a better workplace culture and greater financial results. When it comes to hiring decision-makers, some business strategists believe that it's best to hire people with similar skills to build team cohesion. This practice, however, under most circumstances, causes higher operating costs because your team ends up being heavy in one skillset and weak in others.

It can be challenging for any company, department, or team to establish a productive workplace culture. Productive cultures have a specific recipe, and it takes more than advanced degrees or an ideology to accomplish this. Here is the recipe:

- Have written standards, communicate them clearly, and uphold them.
- Set measurable and attainable goals and objectives, and do not change those without first communicating with your team.
- Create an open flow of communication in and among all departments of the company.
- Ask questions for clarification and provide honest answers.

My personal belief is that people are inherently good, and a company's hiring agent should never take advantage of his/her power during an interview. To ensure this happens, set the standard that all hiring managers must respect hiring guidelines and restrictions. Be professional in your interview questions, and invite the applicant to talk freely as they clarify their answers. When you ask a question, give the applicant time to answer, and really listen while they speak.

Don't look at your notes with your mind on which question to ask next.

The best interviewing and hiring managers have savvy intuition and a thorough understanding of the company culture they are building or want to maintain. They know the deal-breakers: certain behaviors and attitudes, the lack of a certain skill, insufficient experience. Always be mindful that some job seekers are manipulative and will lie through their teeth to get what they want. Be aware of this, and try not to allow yourself to get sucked in by manipulative or fraudulent candidates. Check their references and find out what you can before or after the interview. A wrong hire will result in significant costs and profit leaks. To avoid this, examine the person's ability to do the job.

When you put the emphasis on people, you create a progressive workplace environment.

I'm a proud Canadian, and hockey is my favorite sport. In the game strategy of hockey, the idea is to get the puck to the other end of the rink and into the opposing team's net. You can attempt to do it all by yourself, weaving in and out among the players as you traverse the ice, or you can use your team members to your advantage by passing the puck to the player nearest you, trusting that they will keep the momentum going and move the puck up the ice to the next person. When we break down barriers, we grow as individuals and as a team. When the team has a cohesive strategy, everyone shares in the glory of scoring.

The business environment works much the same way. We hire complementary skills, discuss strategies, and implement goals and objectives; then, we move forward in unison and help others rise. This is the only way to achieve progress and success. During this process, if one person puts up a roadblock and causes profit leaks, you will feel like you're caught in a loop of one step forward, two steps backward, instead of navigating the profit process with forward momentum.

COMMON PROBLEMS WITH EMPLOYEES

Running a business is hard work. Ensuring it is appropriately financed with adequate resources is at the heart of being successful. Keeping the sales funnel full, finding the best person for the job, keeping your people happy and costs under control, and maximizing profits year after year is tough but essential if you plan to be in business for the long term. Hiring the best employees is one of your savviest business decisions, because your workforce stimulates progress and advances the success of your company. It's that simple. Your team members can help you or hurt you. Whatever management sets forth is interpreted and acted on by the workforce. They represent your company to the world. They are on the frontlines, the most customer-facing aspect of your business. They can affect your bottom line more than any other element of your business. Look for employees who will benefit the profit process and not cause profit leaks.

Also, to ensure forward progress in the growth of your business, keep in mind some common problems that inhibit productivity. The following will provide some insight into the fundamental elements of good workers and some red flags to address when analyzing your employees and the work required of them.

INACCURATE JOB DESCRIPTIONS

Although most people understand their overall job responsibilities, there can be challenges conveying the finer details of what is required of them on a daily basis. By taking the time to write a job description in detail and discuss it with the person to whom it applies, you accomplish two things:

- You provide direction to your employees so they have a clear understanding of what you hired them to do. You might suggest that they print out their job description and post it where they will see it every day. This will help them stay focused on what is expected of them.

- Detailed job descriptions help your employees understand their primary and secondary responsibilities. Your role as a manager is to keep everyone focused on department and company goals while holding them accountable for their own actions and interactions. One of the best ways for managers to support employee mindfulness is to set Key Performance Indicators (KPIs) that provide precise information by means of written instructions, milestones, and points of reference for self-assessment.

Most of us have experienced how uncomfortable it is to work in the wrong position for the wrong person. It doesn't bode well for anyone. To create a progressive and productive company culture, you must clearly define each role. This requires finesse, tact, and skill, because personal feelings can interfere with collaborative conversations and ultimately, company-wide decisions. The key is to engage in conversations that lay out and explain expectations in detail.

In your company, if you have a person in the wrong position, you need to take action to remedy this profit leak. You have two options: train them, which is always the best option, or dismiss them. The caveat is to know what you want before you take either action.

LARGE EGOS

The natural tendency for some is to build or expand on their status and level of importance, most often to the detriment of others. This character flaw results in sub-groups of employees working in silos and becoming proprietary of their work and surroundings. "Check your ego at the door" is a saying I often use. Remind your employees that no one within the company is competing against anyone else.

Everyone's mission and goal is for the betterment of the product or service you provide your customers so that you can all succeed. Everyone is a valuable asset; everyone offers something unique. As the business owner or manager, expand your perspective, and your employees will follow your example.

OVERWORKED AND UNDERAPPRECIATED

One of the most valuable lessons I learned early in my career was to appreciate others' efforts. Although I did not always acknowledge it, I always appreciated it. Then, as I grew in a leadership role, it became part of my regular practice not only to appreciate them, but to acknowledge their value and effort. The key is to appreciate in earnest as opposed to giving empty flattery without sincerity. Words are cheap, but actions speak volumes. Don't rely on superficial, empty praise as a low-effort substitute. Demonstrate your true appreciation.

Industry research and anecdotal experience have shown that bad bosses are a major cause of job dissatisfaction. I personally make a point of getting to know every employee—who they are as a unique individual, what they like, what they dislike, their strengths and their weaknesses. When I conduct interviews, I ask a lot of questions to get to the root of who they are and whether we will work together well. This method also reveals whether I will need to change my approach if I hire this person, and how that will affect their job success as well as the company's profit process.

Boosting company and employee morale and valuing your employees are at the forefront of success. Do the managers take the time to know their employees as individual people? If not, the employees may become disgruntled. They may feel like a number and not a valued team member. It is important to know your people, as this helps team members engage on a steady basis. They feel valued because they know their job input matters. As a result, they perform at a higher level.

Everyone has a story. Your job is to encourage your employees' stories and to tell a few of your own, even if they reveal your mistakes. When your employees realize that yes, you are a flawed human being, and you've made plenty of mistakes too, but you've learned from them, they will know that they are valued and their work is important. They'll realize that we are all similar with different experiences. Promote healthy interactions and foster a caring workplace in an atmosphere of sincerity. The positive cultural shift will astound you.

ENTITLEMENT

Entitlement refers to the belief that you deserve something that others do not, and/or special treatment or privilege. This is a very hot topic these days, especially in terms of generational and educational differences, and it has ramifications in the work environment. Entitlement typically presents in a few different forms:

- Young people starting out in their careers who have plenty of energy but limited or no job (or life) experience to draw on, thus making it difficult for them to see alternative perspectives or solutions to problems that arise in the course of conducting business on a daily basis.
- Seasoned employees with ten to twenty years' experience (or more) who intuitively understand alternative ways to achieve results, yet who often lack motivation and energy because they approach their job with a "business as usual" mentality.

Both types of employees are valuable, and the best businesses have a vibrant mix of the two. The key indicator of success is your employees' willingness to look past who they think they are and to honestly assess who they *really* are, their strengths and their limitations. Each employee has a unique skill set and perspective. As a business owner or manager, it is your job to encourage their strengths and thereby minimize their limitations.

This may be a good time to examine the personalities of your staff: their abilities, energy, and willingness to learn. Do they exhibit signs of entitlement? Is that causing a toxic work environment for their coworkers? Would they willingly flip burgers at a staff BBQ, sweep the parking lot, or take out the garbage without being asked? Are they able to do things to gain experience and ability, even if those tasks aren't specified in their job description? Are they aware that there is more to a job than a title? Are they bringing their best knowledge, energy, and attitudes with them to work every day? Are they coachable? Can they identify the responsibilities that link to various roles in your business? Can they make logical choices midstream if a problem occurs—their coworker doesn't show up, or they get confused and need help, or equipment breaks down—or are they expecting someone else to come along to "fix it" or do it for them?

Company culture plays such a huge role in the business environment, and is one of the keys to being profitable. It starts by using finesse, tact, and skill. Use a segment of your staff meetings to define each person's role and responsibilities. The manager's responsibilities should be included as well. When each person understands his/her job description and role within the business, you can have open conversations about their interactions with each other. These can be individual and/or group meetings and discussions that are planned with a specific agenda to keeps things focused and on track. Another excellent way to break down barriers is to engage in random group activities that promote universal engagement, such as in-house training days that help with personal growth, with lunch provided.

These exercises help people relax, which in turn helps them collaborate and obtain a better understanding of their importance in the workplace. These activities should always emphasize that when we collaboratively support each other while working toward achieving company goals, we reap the rewards of creating a better work environment and better relationships with each other. As everyone becomes accustomed to positively supporting their coworkers, they will bring their best to their job. The result is a strong, progressive, and productive work culture.

CHAPTER 5

Management:
What Starts at the Top Trickles Down

Managers are put into position to supervise and lead employees toward achieving company goals with a parallel goal of creating a harmonious work environment. Managers have their own styles and favorite strategies, but there are two distinct types of managers: the micromanager and the laissez-faire manager. In this chapter, we will dissect the specific ways that each style can cause profit leaks. Finally, we will broaden our understanding of good management. First let's examine a fictitious company called Phil's Manufacturing to illustrate the pitfalls of being a micromanager.

MICROMANAGEMENT CASE STUDY - PHIL'S MANUFACTURING

Claude inherited Phil's Manufacturing from his father. Claude was well spoken and portrayed himself as a caring person. He enjoyed his work so much that he spent six days a week at the office and devoted sixteen hours a day to managing his company.

When I first met Claude, I thought he was a fine manager who would be great to work with. We had a similar approach and shared a lot of the same opinions on many issues.

When I came on board as the General Manager for Phil's Manufacturing, my job was three-fold:

1. Enhance the flow and sequencing of processes involved in shop activities and reduce the unmanageable costs associated with present practices.
2. Train the salespeople to price jobs accurately and to limit the number of freebies given to clients.
3. Improve communication between the shop and the main office.

I methodically went through the facility to identify areas of challenge while getting a feel for the cultural makeup of Phil's Manufacturing. After my mini-investigation, I brought my findings to Claude. We strategized various ways to improve shop procedures and increase profit.

I knew the industry was competitive, but the red flags went up when I saw the sales lead, Adrian, giving away free hours and/or applying large credits to clients just to get the jobs. Because of these "deals" with clients, every job required custom-tailored paperwork. Standard protocol was not followed, which caused confusion in every department. Initial data entry was poorly done, and the sequence of information was frequently incorrect, repeated, or missing. It didn't help when Adrian made changes before, during, and after the jobs without telling anyone.

I discovered that these "deals" had been going on for some time. Before this came to light, however, Claude had mentioned to me in private that he'd had enough of this type of free-wheeling decision making from any of his staff, so I was confused as to why this was still happening. I also shared that Adrian had a habit of interfering with the daily productivity of the shop and disrupting the flow of work. Claude stated that Adrian was an important cog in the wheel, and that I need not concern myself with his work practices.

A few days later, I had a conversation with Adrian about doing extra work for clients without charging additional fees. I told him this had to stop unless management approved it beforehand. We discussed a certain client who had received many free hours of work and products. I explained to Adrian how much this cost the company in lost profits and payroll. The conversation with Adrian seemed to go well, and he agreed to be less accommodating of clients' demands. He confirmed that he needed to quote exact work hours, and agreed to seek approval before promising free hours and products to clients.

However, later in the day, he got very abusive and yelled at me for interfering with his work. This surprised me, especially since we'd had such a great conversation earlier that day. What I got out of Adrian's meltdown was that he knew he needed to make some changes, but after he had time to think about it, he silently told himself he did not want to change. He then did what came naturally to him, and when I confronted him, he tried to bully me into allowing him to have his way.

His threatening and abusive attitude required serious reprimand, but when I approached Claude about the incident, I was surprised by his lack of willingness to take corrective action. I was completely baffled by his lack of concern. Claude directed me not to take any action toward reprimanding Adrian. He said to let it go and improve my own skills as General Manager by getting along with Adrian.

I spoke with another employee about the incident and was told, "This is not the first time this type of thing has happened with Adrian and Claude. Just look the other way and don't make comments."

I had a serious concern about the inability of the shop foreman, Ryan, to organize his employees. After several conversations with him and other staff, I realized there were a lot of behind-the-scenes activities going on. For example, Claude would talk to a new client and make a deal, but when the job came in, nobody knew what to do because there weren't any job details, and Claude was often not present during the job review. As a result, we were left guessing as to how to complete the job. When I saw this happen on a regular basis, I understood why Ryan couldn't organize his team. To make matters worse, Claude often made changes to jobs without telling anyone, and then complained about the results.

Claude's unreliable approach to business cost valuable time and resources. He often complained about others not providing clear directions, but he was guilty of that himself. Because he was the owner of the company, he thought a different set of rules applied to him.

The cost associated with this type of behavior is difficult to calculate, as the soft costs to trust are much higher than the hard costs of lost production. The soft costs have a long-term effect on the culture of an operation, which eventually erodes efficiencies and creates cumulative profit leaks.

Phil's Manufacturing was a busy shop, and we couldn't afford mistakes and time delays due to confusing work orders. One of our projects was the installation of a new curtain for a large carwash bay. I had the drawings and understood what needed to be done. I explained my approach and assured Claude of my experience with this type of installation. Nonetheless, he told me exactly what to do as if I were a newbie: how to put it together, where the bolts would go before the concrete was drilled, how the posts were to be secured and supported, and how the curtain should be attached.

Claude and I had a little talk about the value of trusting others and their skills, knowledge, and experience. He relented somewhat, but insisted on being there at every step to pass judgment on my work.

By this time, I had a good understanding of Claude the person and Claude the manager/owner. In many ways, he had created a reactive environment and did not realize it. I was beginning to think he was a control freak who did not trust anyone. He even alluded to the fact that I was not doing my job properly unless I worked as many hours as he did—twelve-hour days six days a week.

On another occasion, Claude and I spoke about the supply chain, the department's poor purchasing habits, and the lack of a tracking system for parts once they were received. We talked about how the parts manager, Betty, was very set in her ways, knew only one way to get things done, and spent much of her time putting out the numerous fires that arose because of her department's poor planning and processing habits. Again, Claude resisted taking action and preferred to focus on the inventory details rather than examining the big picture and the effect these poor practices had on company profits. Based on three major profit leaks—lost parts, no tracking system, and excessive work required to fix mistakes—I estimated the cost of mismanagement in this department to be on average over $500,000 annually.

Claude agreed to discuss these problems with Betty and to brainstorm ways to improve the ineffective practices in the supply chain. During this time, my experience and knowledge of my co-workers was deemed unimportant, and there was little room for anyone's input into those conversations. As a result, there were a lot of complaints, which caused more fires that needed to be put out.

On more than one occasion, Claude told me that the only way to keep things running smoothly was to constantly micromanage the managers. A month later, Claude agreed to hire a new inventory person, Linda. She was tasked with putting an inventory system into place and reporting directly to Betty.

Unfortunately, Linda was not performing to Betty's existing inventory framework, and Betty constantly complained to Claude about Linda. One day Claude stood at my office door and yelled, "You told me you would sort out the parts and inventory problem, and nothing is happening! I want something done right f*****g now!"

I explained that Betty resisted change, and in doing so, she undermined my efforts, and I reminded Claude that he hired me to streamline his business and improve profits. Claude countered that I didn't quite understand the office dynamics, and that it was my job to find a way to get Betty on board. I began to wonder if Claude had his own agenda and didn't truly want to delegate any decision-making power to others. Claude complained about Betty on several occasions and in the same breath suggested I fire Linda. I couldn't figure out why he wasn't taking any action to fix the problems with the supply chain manager. Claude's constant puzzling actions were taking a toll on my loyalty and enthusiasm to bring change to this company. It became increasingly obvious that I wasn't getting all the details. I learned afterward that Betty was "untouchable" because of her history with the company, and I was not the first manager to recognize that as a major problem within the dynamics of Phil's Manufacturing.

Claude had a low opinion of his employees and did not trust them to do their jobs. During our conversations, he often said unkind things about shop employees such as, "They do not have the basic capacity to do their own jobs, let alone what it takes to run a shop, so it makes no sense talking to them about direction or asking for their input." Claude accused them of wasting time, resources, and company profits. He walked around making comments on their work, telling them what to do, when to do it, and how to do it. He was heard to say, "I have been in this industry for twenty-five years, and I know what works and what doesn't." When jobs came back needing corrective work, he put the blame on others.

One of the things that bothered me about Claude was his demanding and demeaning approach he used to "motivate" his employees. There was a common theme among the staff called The

Wrath of Claude: unless you adhered to his every wish, you were subject to his abuse.

He seemed to have favored staff members who received special treatment, but his behavior toward the other employees was often disrespectful. He consistently told staff members that he disliked them, and it well known among the staff who was in his favor and who was not. One employee was extremely skilled, yet he was on the chopping block because Claude did not like him. Concerning another employee, Claude said to me, "As long as he is here, I am going to ride you." I knew what this meant: Claude would make my life miserable until I found a reason to fire the man.

PROFIT LEAKS - PHIL'S MANUFACTURING

Phil's Manufacturing suffered from a stagnant business model, the same one they had used for over twenty-five years. It had probably worked well for some time, but it did not work in the twenty-first century, nor with the current markets. It centered on creating value for one person: the owner. There were limited checks and balances to ensure the company grew responsibly and prospered. With this single-minded approach, input from staff was not necessarily wanted, creativity was undesired, and valuable initiatives were blocked. Claude was not a bad person; he was just stuck in the past. As a result, what could've been a vibrant workplace where innovation and creativity were welcomed and the company prospered had become a stale, stagnant work environment.

Phil's Manufacturing suffered consistent and significant losses in profits due to the lack of proper supply chain protocols. Supply chain is one of the major profit leaks in a manufacturing and service type business. If the right parts are not reasonably priced, do not come in on time, and are not organized or delivered to the technician who needs them to perform the work in a consistent manner, the company bears a cost burden of at least 20 percent. This amount can be much higher if there is no system in place at all.

Claude wanted things to change, but he was not willing to hear what others suggested because he did not want to relinquish control, even though he repeatedly said he did. How was this company supposed to grow without new ideas and new products to meet the demands of new clients? Claude didn't trust his staff to do the work correctly, which resulted in them proving him right by demonstrating poor work habits and creating an apathetic and dispirited workplace culture. He wanted great action but fostered an environment that resulted in poor performance.

Allowing favored staff members to provide extra work at no charge or salespeople to lower the costs for a job is a huge profit leak. If this is happening within your department or company, take heed! Learn from Claude's mistakes. It cost Phil's Manufacturing thousands of dollars to redo job paperwork, juggle priorities, and organize job sequencing—processes that should have been fine-tuned and honed to precision so the business would run like a well-oiled machine.

LAISSEZ-FAIRE MANAGEMENT CASE STUDY - WYATT'S FABRICATION

My experience with this company was pleasant. Wyatt was nearing retirement and had been with Wyatt's Fabrication since it first opened its doors fifteen years prior. I came on board as a possible replacement for Wyatt when he retired. He was communicative, understanding, and had reasonably good standards. These standards had been pre-defined and put in place by the head company located in the United States. Wyatt worked well with people and information, and preserved the integrity and culture of the workplace.

Wyatt usually spent his days in his office attending to paperwork, writing and responding to emails, and making phone calls. He regularly visited the front office to chat with clients and administrative personnel. About once a week, Wyatt walked through the fabrication shop and chatted with the workers. He was a kind man, and maintained good personal relationships with his employees.

He dutifully attended important meetings with the head company and was good at improving existing structures and systems to ensure he wasn't disturbing the flow of activity in the fabrication shop. Workers appreciated his patience with their time and performance. They were mainly left alone to attend to their responsibilities and tasks in their own way. Wyatt wasn't the type of manager that gave too much support and guidance to his workforce. To his way of thinking, they excelled better if he allowed them to do their jobs.

Wyatt was also overly patient with recurring problems. He was inclined to let difficult situations slide with the hope that they would sort themselves out, but they rarely did. Left unchecked, these situations festered and became larger problems. Let's consider one example.

Olivia was the company accountant. She was skilled at maintaining finance records and managing budgets. Wyatt's Fabrication had a paper system in place for finances with ledgers and mark-up sheets. Olivia had gone to Wyatt on several occasions and discussed possible ways to improve the financial system because the filing cabinets had reached capacity two years ago. Each time she brought up this topic, Wyatt calmly replied, "Olivia, please use the system in place. It works, and we don't have a budget for changing the system. I'll buy another filing cabinet for the office."

Olivia was frustrated, so she decided to be innovative and develop her own system. She created an Excel spreadsheet, complete with formulas and a client account tracking system. She went to Wyatt to show him how he could expedite the tracking of clients and monies by using the spreadsheet. Again, and with the goal of keeping the peace, Wyatt calmly told Olivia to "use the systems and processes in place."

Totally frustrated this time, Olivia started looking for another workplace that would appreciate her innovative and creative ideas. She found it and moved on. Wyatt's Fabrication lost a valued employee who might have increased profits by using better systems and more efficient procedures.

Wyatt, being a laissez-faire manager, was overly compromising to his shop employees resulting in staff dictating standard practices. I came in early one morning to get a head start on some inventory issues we were concerned about. As I walked through the shop, I noticed Frank, who had already clocked in, sleeping in the breakroom. I woke him up and asked why he was sleeping. He said he was waiting for Liam, the foreman, to get in so he would know what his tasks were for that day.

I'm sure you're thinking what I thought: not many companies pay their employees to sleep during company time, so why did Wyatt allow this? I told Frank to sweep the shop floor while he waited. When Wyatt came in, I spoke to him about the incident. He shrugged and said, "Okay," and walked away. I never heard anything more about it. Wyatt believed in minimal supervision to the point where the employees ran daily operations. This led to poor job performance, inability to complete projects in a timely manner, and excessive costs.

PROFIT LEAKS - WYATT'S FABRICATION

If Wyatt's Fabrication had a strong, independent workforce, Wyatt's type of laissez-faire management would have worked wonderfully. He would have been the perfect fit for a company with well-honed systems and structures in place and a progressive culture. Instead, because Wyatt was indecisive and detached, and provided little to no direction to his staff, his company was rife with poor employee performance, which led to serious profit leaks. Wyatt did not possess the ability to help people grow in their skills and become the best at their jobs.

Being indecisive is not a productive habit. A successful business trains its employees to focus on quality production and effective delivery. When those are nailed down, everything else falls into place. Wyatt's laissez-faire management style is a prime example of how profits are lost when management does not value accountability and peak performance. This unspoken message filters down through all levels of the organization until ultimately, everyone takes a laissez-faire approach to their job. Standards are jeopardized and assumptions are made when budgets are not balanced, deadlines are unimportant, and schedules are not followed. Maintaining good relationships with employees is one of the primary goals of a manager. Holding workers accountable for their work and their performance is paramount in good management.

Wyatt spent more time managing the business than managing systems and leading the people. His failure to guide his employees resulted in poor performance, which resulted in low profits.

SUMMARY

Today's markets are volatile, and the business climate can change in a day, and sometimes within a few hours depending on larger events in the global economy. In the case of Phil's Manufacturing, Claude needs to build an adaptable business organization, and be willing to use new technology for efficient systems. He would benefit from empowering and enabling his employees to reach company goals through creativity, initiative, and innovation. If Claude taught his staff to make informed and logical decisions about the company and its customers, the employees would focus on quality production and effective delivery rather than fear of recrimination from the owner.

Employees cannot be creative when someone is constantly hovering over them. Claude needs to provide his employees with the right tools, information, and resources, then get out of the way and let his workers experiment, bounce ideas off each other, and be creative in problem solving. Employees can be taught to make good decisions about the company, its products and services, and its customers' needs and wants.

Today's markets require a different approach, namely in the area of communication. If you're a business owner or manager, allow your employees to offer their thoughts and ideas, then listen to them. Share important information so they understand how vital they are to the company's success. Remember, if you constantly tell your employees what to do and how to do it, they will feel squelched and stifled. Rather than putting up with that every day, they will quit their jobs and go work for your competition, and your company will lose a powerful opportunity to be more successful and profitable.

Both management styles—the micromanager (Claude) and the laissez-faire manager (Wyatt)—have positive and negative characteristics. The secret is to develop the strengths of each style, diminish the weaknesses, and to know when to use each style effectively.

Some businesses are able to adopt the agile management style, which allows teams to take ownership of projects and work closely together toward the successful completion of a venture.

According to *Forbes*, "A challenge that many leaders encounter as their businesses grow is finding the management style that works for them and their company." *Forbes* also states that as a business grows, finding the right management style "is necessary to keep talent engaged, teams collaborating efficiently, and business goals achieved."

It's important for managers to have constructive conversations with their employees about cultural factors such as company values, goals, and mission statements. An astute manager conveys the importance of each employee in creating and maintaining a friendly, dynamic work environment to bring those goals to fruition. Although this may be difficult for a micromanager like Claude, who demonstrates an intimidating persona, or Wyatt, who struggles with complacency, both must be the first to make changes before they can adequately direct their employees to follow suit, because leadership starts at the top.

Difficult conversations that involve tough decisions and unpleasant sacrifices might be required to bring about progressive change. Managers should openly encourage the presentation of problems as well as collaborative brainstorming for solutions. At times, managers may need to spend time coaching and training to make necessary changes, to repurpose employees, or to make the tough decision to let someone go if they choose not to change.

My approach to change is self-empowerment, where the managers and staff learn new skills that result in better decisions and a more collaborative and process-oriented environment. This approach is time consuming at the beginning, but once this system is in place, the harmony and profit gains are priceless.

> "Surround yourself with the best people you can find, delegate authority, and don't interfere as long as the policy you've decided upon is carried out."
>
> – Ronald Reagan

CHAPTER 6

Profits or Leaks: Operations, Functions, and Quality Assurance

The Operations component of your business is probably your single largest expense. The reason: it develops a leading relationship with every other department in the company. The operations department keeps all activity flowing in unison.

This year, I mentored senior high school students who were taking a course in entrepreneurship. I introduced them to the concept of Operations when we were discussing the various roles and responsibilities in a business organization. I found it interesting that the students were completely unfamiliar with this concept. They aspired to be the president or CEO, and weren't aware of the makings of a company, how to run a company, or the vital business components that make up a company, all of which the owner, president, and CEO must understand.

In my role as business consultant to various companies, I've discovered that there is some confusion as to what is meant by Operations. Depending on who you speak to and their experience and understanding of running a business, you will hear a few different definitions. Some people think Operations is a function within every department, while others think it's only a fancy word just to give somebody a job title, such as Operations Manager or Head of Operations.

Operations encompasses the people, structures, and systems of your business, as well as how they work together and how the information flows.

The primary responsibility in all companies is Operations. All other processes are considered Functions. It is important to distinguish between Operations and Functions and not allow Functions to direct Operations. When Functions direct Operations, turf wars develop between departments. These disputes disrupt the flow of business, and profits evaporate into thin air. Allowing Functions to direct your Operations is a serious profit leak for any business, and one that can be avoided, but it takes constant vigilance because the shift in power dynamics can happen subtly over time.

To avoid this profit leak, diligence begins with you, the business owner. You are the true Head of Operation within your company, even if you delegate that job to someone else. Your primary responsibility is to govern the entire company, to set the direction of future growth, and to organize workflow. Your job is to designate who is responsible for what and how it will get done.

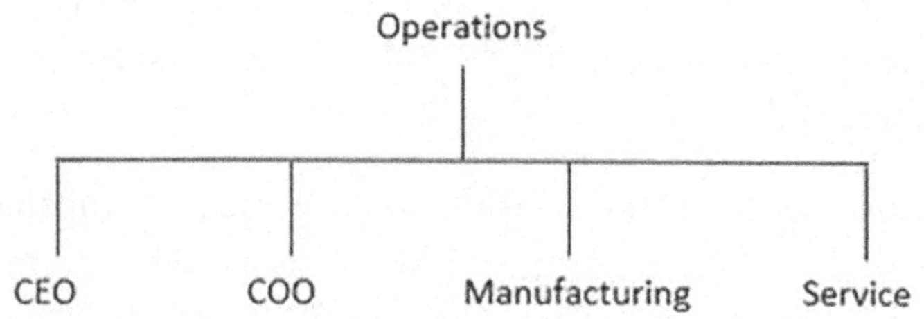

Anyone who owns or runs a company is managing Operations. Being aware of the challenges and relationships among the various hats the CEO wears is important. Many people get confused with this concept, and mistakenly believe that because each Function (department) is different, its operational make-up should be different, and each Function should determine the best way to achieve its goals. This cannot be further from the truth. Two companies may look different from the outside, but their responsibilities are the same on the inside: the primary source of Operations is the CEO.

As with the CEO, the Manufacturing and Service Departments have many vital roles within the company and get much of the work done. Their primary responsibility is to determine, plan, and bring together the final product of whatever the other departments imagine, design, and create.

The Sales Department also serves as a Function (component) within any business, not an Operation. Their primary role is to work with clients and determine what the clients require, obtain input from all other departments, and offer a sellable option. The Sales Department supports and feeds the Manufacturing and Production Departments.

The Engineering Department also provides a vital Function. This department provides technical advice, data, and specifications in the form of technical drawings to the Manufacturing Department.

The Supply Chain Department purchases inventory, manages supply, tracks shipments, and delivers products. Although this could be considered an Operation in part, it is a secondary activity and therefore defined as a Function. Their primary responsibility is to provide support services by purchasing, storing, and delivering products to other departments within the company.

The Accounting Department tracks, monitors, and manages the company's money, but it also projects company cash flow and shares financial information with all departments. This function enables department managers to make valued and responsible decisions regarding the costs involved in running their departments.

SERVICE AND MANUFACTURING CASE STUDY – LUKE'S INDUSTRIAL

The following is a real example of a service and manufacturing company with poor quality assurance and inefficiencies in Operations. We will examine the negative effects that departmental decisions had on the overall success of the company.

Luke's Industrial was new to the North American market. The company boasted an innovative high-tech drilling rig that would lower costs and increase efficiency when drilling wells. The owner, Luke, was looking to hire a new manager to sort out the service and production department challenges. That's where I came in.

I was hired as manager of both departments. I followed my usual process and did a thorough company analysis. Almost immediately, I noticed the complete lack of organizational structure in the service department. The service people dictated all activity. There was no production schedule to formalize when certain tasks had to be started, completed, and delivered. Service techs chose the jobs they wanted to work on and decided when they wanted to work on them. Some service techs disappeared for days—they literally did not show up for work, and no one seemed concerned about it. Furthermore, the Service and Manufacturing Departments were pulling from the same employee base, which put additional strains on the facility's productivity and made it difficult to schedule the work with any form of consistency.

Inventory control was nonexistent. Techs grabbed whatever they wanted from inventory and did not record what they took. This resulted in job costs being inaccurate, which necessitated large write-downs, all because inventory was not managed properly.

The Engineering Department was top-heavy with management, and too many decisions were stalled due to disagreement and lack of a cohesive vision. It was disheartening and frustrating to see inaccurate drawings used to build technical equipment. This caused delays and cost overruns because the drawings had to be returned to Engineering and reworked. Because of the rework, finger pointing and blaming ensued. There were arguments among and between the various departments, and ill feelings all around. It was common to hear that engineers changed things at the last minute because they had made errors in the initial project design. Sometimes the engineers made changes during last-minute phone calls with the customer after the drawings were supposedly complete and ready for production.

The Manufacturing Department had difficulty building one drilling rig at a time because some people liked the current assembly method and did not respond well to change. There were ongoing scheduling conflicts and constant bickering over the level of importance between trades and engineering, which exacerbated the problem. The department was plagued by indecision, skewed priorities, and slacking off. Workers dictated their activities and hours. This drove up costs astronomically. Jobs that should have taken 500 hours to complete took 1000 hours.

The Operations Manager who directed the Engineering Department had a bad habit of micromanaging people, and was very defensive when asked questions about his processes and procedures. He seemed more focused on doing the President's job, and he interfered in the workflow of other departments rather than developing a strong and unified Engineering Department with quality standards and output. He felt compelled to tell capable people how to do their jobs, and complained to the President when things did not go as he wanted.

Communication throughout the company was sporadic from the top down. Important information such as delivery dates or customer delays was not shared with those directly involved in meeting deadlines. As some employees said at the time, "It's not important for others to know," even though it was crucially important.

Luke's Industrial had one client who tended to be unrealistic in his expectations of how he wanted his work done and when he wanted it completed. This same client couldn't clearly articulate what he wanted done, however, and he constantly complained about the poor service to the point of being abusive. Additionally, he had a history of not paying his bills.

As time went on, I became aware of another problem with this client. For a period of seven consecutive Fridays, this company called me late Friday afternoon to report that a machine had broken down, and they needed a service tech to come out right away to fix it. They probably knew it would take some time for me to contact and obtain a service tech to drive three hours to their facility. Even with a quick response, I had no way of knowing whether the tech would have the necessary tools to do the repair.

I tried to figure out why this company had so many breakdowns. After thinking about it, I realized they always called late Friday afternoon with a breakdown, meaning if we didn't fix it right away, it caused them down time. Added to that was another thought: Did the breakdowns really occur late in the day on Fridays, or did they occur earlier in the week, but the client waited to complain about it by calling us to complain late on Friday afternoons?

PROFIT LEAKS – LUKE'S INDUSTRIAL

"All over the map" organization results in cost escalations and continual employee and client problems. The poor practices in Luke's facility and the lack of quality assurance in client operations resulted in frustrating cash flow problems for Luke's Industrial.

The operating costs for Luke's Industrial had increased by roughly 25 percent due to poor management, so if we use $3 million as a base, that meant a profit loss of $750,000.

To significantly reduce these profit leaks, we first needed to address the management component, because management set the direction of the company. Whatever management decides will dictate the course of action that is taken by every department.

Organizational instability in any business has the potential to translate into huge profit leaks. The lack of direction from a manager/leader creates massive confusion among staff. As we saw with Luke's Industrial, when non-management staff had the freedom to make decisions beyond the scope of their position, they did just that, at a detrimental cost to the business. Workers did not really know their specific goals and objectives, so they tended to work longer hours to complete their work once they knew what had to be done because by that point, they were scrambling to make the deadline. As a result, the dollar value of the employees' work declined and production became less efficient. Adding to the problem was that different departments were pulling workers from the same pool, which led to more confusion. The workers never really knew what they were supposed to do or where they were supposed to be stationed.

These bad habits were so deeply entrenched that the resulting profit leaks seemed like normal operating procedure, but Luke knew he had some major problems to resolve. He didn't know how, but he knew he wanted and needed to make some radical changes in the way he ran his business.

Luke's Industrial made some hard decisions with the goal of getting the Service Department to run more efficiently. Management included various departments in the decision-making sessions to allow them the opportunity to voice their thoughts early in the planning process.

Luke's Industrial needed to streamline their daily operations in five key areas:

- Clearly define and optimize employee roles and responsibilities
- Create logical schedules
- Refine planning protocols
- Create comprehensive schedules for the Service and Production departments

Once this was done, management introduced the restructured roles and responsibilities of each employee, trained them to understand their role within the company, guided them to respect and follow schedules and protocols, and to be accountable for their work within those protocols. Additionally, I recommended that Luke's Industrial offer a Trades Apprenticeship program to properly train new employees while they job-shadowed experienced employees.

To maintain high standards of quality and excellence, I recommended that management create interdepartmental collaborative environments through regular "toolbox" meetings to help everyone understand their coworkers' perspectives and challenges. A toolbox meeting is a mini-meeting of the shop and management personnel to discuss direction and answer questions. These meetings can also serve as planning sessions. Ask everyone to bring their best cost estimates for production—the cost of resources, people, materials, and staff hours—so that as a team, you can strategize how to meet deadlines and stay on budget.

Another profit leak at Luke's Industrial was inconsistent inventory standards. This led to inaccurate job costs and excessive write-downs at the end of each job, which had a cumulative effect by year-end.

The Accounting Department needs to know what has been procured, what is being used, and what costs are associated with the operations involved. Additionally, a successful company needs to have a controlled protocol for tracking inventory.

As a rule, engineers develop new products. They use the principles of science and complex math calculations to produce the perfect specifications for resources to be developed. From these calculations, it is expected that engineers bring forth exact measurements and specifications for new pieces of equipment. When forced to have drawings reworked, the company suffers lost time and lost wages. Added to this issue is the all-too-common scenario where an engineer "needs just a few more hours to get it right." Added hours to "get it right" means cost overruns for the company, a big profit leak.

When the Engineering Department is top heavy, miscommunications and turf wars arise because too many decisions are stalled at the top. This was a major problem at Luke's Industrial. I recommended that top management fix this profit leak by examining and restructuring the roles and responsibilities of the Engineering Department's managerial team. Another solution was to hire an Engineering Manager to run the daily functions of the department.

We all know that when clients complain about the service they receive, something is wrong somewhere. In this case, Luke's Industrial needed to work with the client by changing habits and tendencies and creating respectful service protocols. The first step was to have a meeting with the client to outline the roles, responsibilities, and expectations for Luke's Industrial and for the client.

During this meeting, Luke offered the client the opportunity to keep his costs in check by drafting a maintenance service contract, which enabled him to schedule dedicated technicians to work exclusively for that client. This would keep costs down and maintain the equipment at an optimal performance level. Luke would ensure that his technicians had the proper tools to do their jobs and arrange for them to follow proper training to use the tools correctly. A box of spare parts could stay at the client's facility to expedite the repair of common problematic equipment parts. As an added bonus, Luke could establish higher client standards by training machine operators to use equipment properly and within its capabilities.

Additionally, and with the goal of significantly reducing last-minute service calls, Luke's Industrial implemented a policy that informed the client that when a bill was 30 days overdue, the client was required to provide a service deposit via wire transfer into Luke's Industrial's bank account before a service tech was sent to the site. Furthermore, if any tech accumulated 44 hours in any given week, overtime pay would be in effect for the entire service trip for that week.

Not everything at Luke's Industrial turned out well after the implemented changes. The President decided to expand the Operations Manager position to include an Engineering Department Manager and an Operations Manager. This had a negative effect on employee morale because the Operation Manager continued to micromanage entire departments and those around him, which created frustration and lowered the level of accuracy and engagement in the Engineering Department.

A new Engineering Manager was hired, who turned out to be a true asset to the company. He raised the level of accuracy throughout the Engineering Department, but he was restricted in his attempts to make improvement across the board because of the Operations Manager's interference.

The Plant Manager was laid off because he wasn't performing and meeting target expectations. A new Service Manager/Plant Manager was hired to manage all shop and production activity. This new hire resulted in a reduction in operating costs, a significant reduction in errors, and people taking greater pride in their work. Labor hours went down. Satisfaction and production increased dramatically.

A staged manufacturing schedule was implemented. This increased employee engagement resulting in productivity doubling and a significant reduction (30 percent) in the hours needed to build a given product.

Troublesome client habits changed when the management at Luke's Industrial instituted a preventive maintenance program. Late or after-hours calls dropped by 90 percent, while quality and on-time service flourished. Clients accepted the cash-in-advance policy and paid their bills on time because they understood that this was an established guideline that would not be altered for any reason.

The following year, the company completely changed their philosophy and direction. They tried to save some money by eliminating the Service/Plant Manager position. This major change in the business organization proved to be disastrous. A short time later, the President lost his job, the Operations Manager took over, productivity levels decreased, employees quit, costs increased, and the company eventually went broke.

SUMMARY

I've mentioned that I enjoy the game of hockey. Some years back, I played in an adult recreational league that had some problems brewing beneath the surface. Players and teams were frustrated with league administration because they consistently ran short on cash, and at the end of every year, they were in the negative, mainly because team members did not pay their ice and league fees, and there was no mechanism in place to collect owed money other than guilt. It got so bad that the league faced closure if they did not stay current with their bills.

Two of us took on the roles of President and Treasurer to put some structure into the league. Our first order of business was to calculate our monthly and annual expenses, and to determine team fees. Next, the league executive created a constitution that outlined rules, responsibilities, and consequences for poor decisions, and to establish the financial obligations of each team. In short, if you did not pay, you did not play. Even though there were a lot of heated discussions about these changes, consensus was met, and the league never went into the red again. This scenario clearly demonstrates that poorly managed operations will only cause frustration and profit leaks whether in business or recreation.

CHAPTER 7

Profit Process and Leaks Case Studies

A company that operates successfully generates a profit and increases the value of the business. On any given day, profit leaks can happen when workers don't pay attention to detail, fail to communicate, and/or forget to follow standard protocols. In this chapter, we will consider several scenarios of common problem areas where profit leaks occur in businesses of all types.

Greg is new to the Sales/Marketing Department, and is eager to prove himself, but sometimes his enthusiasm gets ahead of his inexperience. He forges ahead with a new deal, not realizing it is an unprofitable one, and without properly understanding and discussing with the team all the protocols required to bring that sale to fruition. Instead of adding to the company's bottom line, Greg has inadvertently created a profit leak. To remedy this problem, a one-on-one meeting is scheduled with Greg to discuss his thoughts and to reiterate the importance of involving all the stakeholders involved before proceeding. During this meeting, management emphasizes that this procedure might take an extra hour or more at the beginning of the process, but in the long run, hours of rework will be eliminated.

A general rule concerning rework is that it takes between 2.5 and 3.5 hours of extra time to fix something as opposed to taking a little extra time at the beginning of your planning cycle.

If you are short-staffed and decide to place someone from another department into a sales position without adequate training, this may provide a quick fix, but it may also create a profit leak for your company due to documentation errors, inaccurate quotes, and inexperienced salespeople promising more than they can deliver. Diversifying your workforce (cross-training) creates a win-win situation for the employees and the company, provided your employees are not thrown into a situation without ample training and realistic expectations. The latter part is where many business owners and managers fail. If we push too hard, cross-training will feel like a chore to our employees and not an opportunity to increase their knowledge base.

Mei has been in the HR department of Archie's Environmental Company for twenty years. She has been responsible for the recruitment and selection of the most productive candidates they've hired. Within the past few years, however, the nature of the HR department has changed, and she has been tasked with more responsibilities. She now manages training and compliance, health and safety, and the occasional onboarding of contract workers during busy seasons. The addition of more responsibilities strains Mei's ability to complete her work effectively, and what might have seemed like a good decision at first—to pile more work on the most competent employee in the department—has become a profit leak because there are not enough hours in the day for her to complete her work accurately and effectively. This causes unintentional delays, unfinished work, incorrect work, and the need for do-overs.

Does this scenario sound familiar? Take a moment to consider all the people and departments within your business. Are your key employees being challenged with too many responsibilities? Are expectations too high or unreasonable? These are a few questions you need to ask yourself before piling on the workload. Valuing your employees' work- and home-life balance is essential to maintaining a healthy work culture.

It is also important to consider the well-being and mental health of your employees. Will your decisions affect workforce retention? When decisions are discussed, made, or implemented, who is responsible for dealing with the ethical conduct of staff? These issues tend to be hidden or ignored, and they emerge later as profit leaks.

Profit leaks can happen in all sorts of subtle ways in today's business environment. One of the most common areas has to do with technology. A few years back, my wife was giving a computer workshop to the teachers of a large urban school board when suddenly, they heard the familiar sound of a printer buzzing and clicking in the adjacent room. My wife asked if anyone had sent something to the printer, as they were the only ones in the building at the time.

"No," came the collective response.

She walked over to the printer, picked up the printed page sent from an outside source, and read aloud, "If this is the best your school board can do for security measures, I'd say you're in big trouble!"

As it turned out, the printed message came from a previous IT employee who now owned his own IT business. The school board made the wise choice to hire him on a contract basis to ensure their cyber security stayed up-to-date. This solution didn't take away all the profit leaks in terms of technology within the school system, but it was comforting to know that someone with knowledge of cyberspace gaps was working in the background to keep everyone's data safe and the system up and running.

As we discussed earlier in the book, the profit process is a continuum that involves all the departments within the business, starting with sales and overseen by management. The following graphic accurately conveys the importance of this interconnectedness:

As depicted at the top of this graphic, the profit process begins and ends with the customer and his or her order for a product or service from your business. As each order travels throughout the various departments that make up the typical business, profit leaks can happen. Let's examine some case studies to illustrate this.

ADMINISTRATION PROFIT LEAKS – PATTI'S ENGINEERING CASE STUDY

During Nancy's interview for the job of administrative assistant at Patti's Engineering, she states that she is good with a computer, she's professional, and she pays attention to detail. She's appropriately dressed for the interview and seems comfortable discussing various topics at length, so she is hired as an administrative assistant. Nancy arrives on her first day of work dressed in a see-through white blouse over a black bra, short leather skirt, and high heels. Her boss dictates an important letter for Nancy to type and save in the computer network, and he asks that a copy be sent to the regional manager. Later in the morning, he notes that Nancy seems to be on a personal phone call. Afterwards, he sees her looking at houses on a real estate website. The next day, Nancy calls in sick. The following day, when it comes time to create, file, and share the important letter, Nancy's boss notices several errors in the document. Nancy struggles with work-related computer tasks and makes excuses to justify her lack of ability. This is a very costly hire, and corrective action needs to be taken immediately to replace Nancy with someone who has the required skills to meet the job requirements.

As a side note, when I hire staff, I usually have some sort of test for the applicant to take to see if their skills match what they state on their resume or during the interview.

Time is money. Problems in the area of time management such as excessive time spent chatting, scrolling through social media, being consistently late, and chronic absenteeism are not beneficial to the employee or to the team they support.

Mistakes and problem behaviors can occur in any business, but consistent, repeated mistakes need immediate attention. If you wait too long before addressing the errors, you risk having these inferior practices become the acceptable norm throughout your administrative support and departmental processes. Quick and professional corrective actions are required to ensure adequate standards are maintained.

Our workdays are busy. Finding information quickly is crucial. Organized filing systems, both paper and electronic, are essential to the success of your business. For electronic files, if you cannot find a document within three clicks of your mouse, there are usually some problems with your digital file management system. It is either so complex that it would be beneficial to have it streamlined, or so unorganized that it needs to be rebuilt. Either way, the organizational structure of the system needs to be improved.

Physical files must be legible, have a relevant label, and adhere to a consistent filing system throughout the office so that everyone can find what they are looking for. The file system must also tell a chronological story, with the most recent files and documents in the front part of each folder.

Many businesses tend to be double-minded when it comes to retaining staff, even when certain employees have limited skills. Some companies believe that every skill can be taught, and others have a fear that if they spend profits on training, their employees will leave and find better positions elsewhere. As a result of this fear, training budgets are cut or eliminated altogether. A question that I always ask myself, and anyone responsible for hiring, is this: Are you hiring the person for the job, or are you attempting to make the job fit the person? Are you interested in the applicant's abilities and accomplishments, or are you blindsided by an attractive resume filled with impressive job titles and corporate jargon?

Often, I have found that companies that have rigid or restrictive training beliefs usually have an internal culture/trust problem. Of course, I believe that companies and people need to invest in each other to be successful. However, to solve skill-level problems, senior management needs to be clear as to exactly what they want; then, they need to communicate this vision, seek feedback, and practice what they preach. That last part is most important: always seek feedback and walk the talk with sincerity and earnestness, not as a box-ticking exercise. When you are genuinely interested in the thoughts, ideas, and suggestions of your employees, you create alignment with company goals. As a result, you will plug profit leaks, and your business will grow exponentially.

Administrative support personnel hold a key position in any company, though they are often perceived as being at the lower rung. They manage the daily details that affect whether your business runs smoothly or goes off the rails. They can set the pace and raise or lower operational standards based on their decisions and whims if not provided with meaningful direction and understanding. For this reason, those who are drawn to administrative positions often have their own ideas of what they want to do rather than focusing on what they need to do. This results in serious consequences if not managed properly. As the business owner, you should provide clear direction and set specific parameters within which your administrative support personnel have autonomy. Establish written guidelines that detail which tasks need the approval of a manager and which tasks are within the employee's purview.

PROCUREMENT AND INVENTORY DEPARTMENTS (SUPPLY CHAIN)

This department or person ensures that the proper components are purchased and assigned to the exact job that needs them (the first time). The Procurement and Inventory departments have a continual, interfacing relationship with Quality Assurance, Service/Production, and Accounting.

Supply chain has the potential to make you go broke in a hurry if the process is not managed properly. For starters, the procurement and inventory departments have completely different functions that tend to be ignored or misunderstood. Too often, this department or people do not mesh with production or the service the company provides. These poor practices result in exasperating problems and profit leaks.

The following is an example: Lucinda found a great deal and purchased 100 60-watt light bulbs at a cost of $1.00 each, for a total cost of $100.00. However, after the purchase, Lucinda learned that the owner had planned to replace all the incandescent bulbs in the office with LED lights. You cannot fault a person for taking initiative, but sometimes misguided initiative comes at a cost. In this case, communication between the purchaser and a supervisor caused a profit leak.

Were Y products requested, but X products were purchased because they were cheaper, or was it because the purchase order request was unclear? Is the purchasing decision based solely on price and not on quality or service? These issues are more prevalent in larger organizations, but the takeaway is that requests should be clear and directions followed according to company standards.

PROCUREMENT PROFIT LEAKS – ROBERT'S MECHANICAL CASE STUDY

The following is an example of how profit leaks in the procurement department can have a ripple effect throughout the entire operation. On Monday, Robert's Mechanical learns that a fleet of 40 city work trucks will be coming in for maintenance starting Wednesday with a two-day turn around. Kyle, the procurement officer, is charged with purchasing a new hoist to accommodate the repairs. The hoist the service team requested cost $6000 with next day delivery. Kyle finds a vendor that offers a hoist for $4500 with three-day delivery, so he orders that hoist thinking he has saved the company $1500, even though he knows the hoist won't arrive until Thursday. The fleet trucks start arriving on Wednesday, three at a time. The hoist arrives late on Thursday and still needs to be installed, which means that overtime will need to be paid for the installation to happen so the hoist can be operational by late Friday. Because of the late arrival of the hoist, the first round of trucks will not be delivered on time, and that will have a negative impact on the delivery of all the trucks that need to be serviced.

This case study is a perfect example of how the procurement department must be in sync with the service department. Because of his inexperience, Kyle didn't realize that the results of his decision cost the company an extra $2000 in lost production (2 days X $1000/day) and an extra $450 for overtime and after-hours installation. That's a profit leak of $2450 to the company, not to mention the frustration within and between departments and the damage to the company's reputation, and the likelihood that the city might cancel its contract with Robert's Mechanical.

Make sure a follow-up procedure is in place for all products ordered and delivered. If there is no follow-up, parts may not arrive on time, production time is lost, and the workflow in several departments will stall because they can't move on with their part of the work. This causes a domino effect of multiple and successive profit leaks happening at the same time.

INVENTORY PROFIT LEAKS – PAYTON'S MILLWRIGHT CASE STUDY

Payton's Millwright dismantles, repairs and reassembles machinery for large construction and mechanical firms. The amount of necessary parts needed at any given time is massive. Simon is the new inventory manager. As he's checking inventory, he sees Martha, a welder, take a box of welding rods from the supply room and return to her station. He raises his eyebrows when he notices Ben take a few pillow block bearings to his station. After coffee, he sees Patrick take a small shaft and another pillow block bearing to his station. Simon now realizes why he's been hired as inventory manager. Parts are not being tracked or recorded, and there are significant losses in inventory and inaccurate job costs as a result. The owner at Payton's Millwright had to write off a million dollars in lost inventory at year-end.

It stands to reason that parts need to be tracked and/or signed out to a job when they are taken or removed from the shelf. One of the best ways to ensure this is to hire a supply clerk who is responsible for picking parts and filling the orders of the day first thing in the morning or a few hours before they are needed. The supply clerk or an assistant then delivers the parts to the job location and technician. This practice requires an accurate bill of materials, which takes us back to one of the foundational aspects of running a successful business: plan and prepare for the cycle of events at the beginning of the profit process when the sale is made, and there will be fewer problems and profit leaks when the product or service is delivered.

The cost of supplies is a major expense that accumulates capital investment quickly without careful management. Tailor your inventory management system to meet the needs of your business, put that system in place, and follow it, even if your employees complain that it is not what they want to do or are used to doing. Having said this, there are times when a good conversation is beneficial, and revamping your systems is the right action to take.

Your stock costs need to be captured if you want to balance your budget. The larger your facility/company, the more robust system you need to track all costs. Make sure stock counts are done regularly. Every part should have a designated wholesale cost to the company and an associated retail price to the customer.

A good practice is to have the inventory clerk pull products to fill orders as the orders cross their desk. Another is to have workers sign out parts and products at appropriate intervals, such as when they need a specific part. For consumable items such as nuts and bolts, cotter pins and other such items, a third party can come in to count and fill the bins. The costs can be totaled at the end of each month and assigned to the jobs in progress.

Inventory should be tracked and recorded when it goes on the shelf or is taken off the shelf. Here are a couple of examples I've seen that may help plug profit leaks in inventory:

1. Part numbers are assigned and entered into the system when each part is received and stored.
2. All parts pulled for a certain job are assigned a job number, and then delivered. If inventory is stocking parts for a job, the area in which the parts are kept is clearly labeled on the side of the box or on the shelf.

The following are two different types of numbering systems you can use for tracking inventory:

1. Smart – These part numbers have specific characters so you can categorize and identify the part quickly. For example, suppose the product is a box that holds 12 whole-wheat sugar-covered donuts. The PN could be 1B12HWSD.
2. Lineal – These part numbers are assigned according to a pre-generated list of sequential numbers. Each part generated would get the next number in line. For this example, I used a six-number sequence.
   - 123456 – paper napkins
   - 123457 – plastic forks
   - 123458 – small coffee cups
   - 123459 – a 5–gallon jug of cooking oil

Lastly, purchasers should receive product-specific training to understand the consequences that poor purchasing practices have on increased inventory costs. It is essential that they understand delivery dates so they know when and how much inventory is required and what the turnover rates are. When ordering anything, the purchaser should ask specific clarification questions, such as, "When do you need the item?" and "Is this a hard delivery date, or is there some flexibility on when the inventory needs to be on site?" This is vital to obtaining the right part(s) on time for the project. Product needs to be picked on time, counted, and checked on a monthly and an annual basis.

SERVICE AND PRODUCTION DEPARTMENTS

The service and production departments are responsible to deliver the product to the external customer, which makes them the primary internal customer to all other departments. They must create a job plan and ensure that all the information is correct and the key people have been informed; the service or product is accurate and delivered on time; and lastly, that all the paperwork and data is collected and filed properly. If any of these steps are not completed, the costs to any given job will grow exponentially. Often as we rush to finish jobs, we skip steps and put off certain aspects of the job thinking that we will complete it faster. This is not the case, as skipping steps out of convenience, with the promise to do them later, often ends up taking more time and costing more.

PRODUCTION PROFIT LEAKS – GRETA'S KITCHEN GADGETS CASE STUDY

Greta's Kitchen Gadgets is renowned for its ingenious kitchen gadgetry. During a "meeting of the minds" production meeting, the team examines the manufacture of a new Frog Face Toaster. Each toaster is intended to sell for $100 according to market analysis. Sales projections are to sell ten toasters per month. Based on a departmental analysis, material and labor costs will be $50 to produce each toaster. These costs are based on building a three-month supply. Corporate wants to make a 30 percent profit on each toaster within three months. This leaves you with $20 per toaster for engineering, sales, and marketing. With such high production costs and low profit margins, you will suffer tremendous profits leaks, and you will not earn enough to make more toasters, let alone sell them at a profit.

The delivery of the product or service is where the money is made or lost. First you need to look at the role each department and/or person plays in creating profits and reduce costs.

As an example, the sales department creates revenue. They do not create profit, but they can reduce costs by selling more products. This large demand enables the production department to manufacture in larger runs. This in turn makes it less expensive to make the toaster, and clients may receive volume discounts depending on the profit margin the company instructs the sales department to factor into the price of the toaster. If this is done, the profit per piece will shrink and revenue will increase.

Understand your costs, and be cautious and thoughtful when offering volume discounts, as they may bleed you dry.

The engineering department does not create profit, which means they are a cost center. For engineering to create a complete concept and drawing of the Frog Face Toaster, their hours would, at a minimum, triple to complete the last 20 percent of detail on these drawings, and may need to be adjusted when the toasters are assembled due to manufacturing layout and requirements. If this approach is taken, development costs can escalate to the point where the product is no longer profitable to make.

The manufacturing department creates profit through the reduction of costs by adding efficient processes and incorporating planning at the beginning of the manufacturing process. They can also take the last 10 to 20 percent of time that the engineering department would like to have to complete the drawing, calculate the alterations, apply them at the time of manufacturing, and come out with a functional toaster in a fraction of the time. This practice can save time on production and avoid the back-and-forth between departments.

If we fail to plan, timely execution is impossible. Taking the necessary time at the beginning of any process to plan the entire project will produce consistent results in a product or service that flows seamlessly through the entire production cycle. This practice also reduces worker stress, since most issues are resolved in the planning stages.

ACCOUNTING DEPARTMENT

Although accounting has an administrative component, this department keeps a close eye on costs and revenue. Accounting ensures that all costs are accurately documented and balanced. Accounting provides a service to all other departments, as they are tasked with gathering and sharing pertinent information so the relevant parties are kept up to date on the status of production at every stage, with all costs tracked so you know in detail where the costs came from, thus making it easier to determine the make-up of your profit margins. Without accurate numbers, you are simply guessing at your actual costs, meaning where your money is spent and how the revenues are created. If this practice is not followed, you might end up spending the same dollar twice.

But the accounting department can also be another area that creates profit leaks. In their quest to ensure the balance of numbers, human interaction suffers, as the people and location of this department are usually somewhat isolated from everyone else. There are methods to increase interaction, such as having managers do some cost accounting themselves.

ACCOUNTING PROFIT LEAKS – IRONSIDE MANUFACTURING CASE STUDY

I worked as the general manager for a company we'll call Ironside Manufacturing. The chief financial officer was very controlling toward his managers concerning how he wanted them to run their departments. Ironside was consistently in the red. One of the problems that I found when I took over running the manufacturing facility was the CFO spent a lot of time dictating how manufacturing should run the facility and departments, down to the smallest detail: what they should buy, how they should buy it, and when they should have it delivered.

Although the CFO worked in a different city, his actions played a huge role as to why the facility was consistently at a deficit and had a large turnover. The historical data for this company was that during a six-month timeframe, 75 percent of the staff was new. For this reason, the company had difficulty fine-tuning their processes so that everything ran like a well-oiled machine. Every couple of weeks someone left or was hired on. The rule of thumb in business is that every time a person is hired or leaves, it costs the company one year of wages, but I believe it is more like three months of wages and three months of lost value. The calculation considers hiring costs, administration, training, and re-training costs, negative energy and inability to build the business.

When we put hard numbers to employee turnover and the profit leak it causes, for a person that makes $60,000 a year, one-quarter of that (three months' wages) is $15,000 in lost wages and another $15,000 in lost productivity. In this instance, the staff count was 15 with a turnover rate of 75 percent, which means 11 people rotated through the company within 6 months. Assuming all these people earned an equivalent income, the company lost $165,000 in hard costs and another $165,000 in lost productivity, totaling a loss in profits of $330,000. By addressing this one area in Ironside Manufacturing, the savings were so significant that instead of closing their doors, they expanded their operation. As it turned out, when this CFO was removed, a lot of the internal problems disappeared.

Accounting departments are a bit invisible at the best of times because people and companies are reluctant to reveal and explain the numbers. Although there is some understanding as to why this happens, the practice creates a problem for department heads that require accurate data in a timely manner that is meaningful to them, not just the accountant. Trust is required on both sides. Does the accountant not want to show the numbers because they do not want to explain them, or is the department head just being nosy? People are hired to perform duties, and withholding relative information is destructive to the business. If the information is being used for self-serving purposes or nefarious reasons, the problem should be addressed immediately.

Time, labor, and materials need to be tracked daily by the person who schedules and/or organizes production or services. If defined project costs are not being tracked and summarized, the responsible person needs to be questioned as to why. You might get an answer contradictory to productive direction. It is important to ask the hard questions, but also to provide answers and take appropriate action.

Bookkeepers and accountants are essential to the well-being of a company. Numbers with meaning and relevance are more easily understood and therefore regarded as important. The employees in your Accounting Department need to understand the full scope of the people, products, and/or services your company provides, and to provide relevant numbers to each department every month.

The accounting department provides all the financial information after everything is completed. They are not involved in weekly production or service activities, and are not usually required to provide updates and insights along the way. However, integration of the accounting department into production and service activities saves time and increases profits through a better understanding of the challenges that occur during the revenue-making process.

DELIVERY, SHIPPING AND RECEIVING

The last part of the profit process is delivery/shipping/receiving. Without the ability to deliver what you promised, you will be out of business in no time. The conduit for this is the shipping and receiving department. They ensure that what has been ordered is completed, documented, and delivered to the client without delay.

DELIVERY/SHIPPING PROFIT LEAKS – HARVEY'S AUTOMOTIVE CASE STUDY

Ed is the purchaser for Harvey's Automotive. Ed received a request to rush-order an intake valve for a vehicle that was to be repaired. Unfortunately, the part they had on hand was the wrong one. When Ed ordered the new intake valve, he put the order on the board expecting the driver, Alex, to look at the board at a certain time and pick up the necessary part. Alex did look at the board and went out to pick up the part. He decided that since he was out picking up parts anyway, he would pick up a couple of other parts on the list at the same time, but he would grab those parts first because they were closer according to the route he mapped out. Unfortunately, by the time he got to the supply store to pick up the rush part, the store was closed

As a result, the night shift at Ed's Automotive was not able to finish the installation, which meant they had nothing to work on for eight hours. The only alternative was to phone the parts store and ask them to open the doors after hours so Ed's Automotive could obtain the necessary part, and the store agreed, but the mere act of opening after hours had a cost of $500 attached to it. Why? Because they had to pay an employee overtime to drive to the store, open it, turn on the lights, turn on the cash register, recount the money at the end of that brief shift, and update the inventory management system before closing the store again for the night.

The shift supervisor made a deal with the parts store to pay the extra costs to get the parts so the night shift could complete the job. If the supervisor and store had not come to an agreement, an additional $10,000 would have been added to the job in lost productivity. As it turned out, an extra cost of $2000 was added to the job in lost productivity and extra parts costs. What is not calculated here is the soft cost of morale and trust.

There are three profit leaks in this example. The first one is that Ed originally ordered the wrong part. He may have been misinformed, he may have been sloppy with his identification of the part for that certain vehicle, or he may have assumed he knew what was needed. He probably felt he had the right part and everything was good to go.

A second profit leak is that Ed's order form containing the pertinent information was not hand delivered to Alex, and the rush order was not attached to the driver's note. This is a perfect example of why communication is so important.

The third profit leak is that Alex didn't follow the directions on the order form because he didn't read the instructions until after he got to the parts store to pick up the intake valve. He missed the most important information, that he needed to pick up the intake valve first, deliver it to Harvey's Automotive, and then carry on with his day.

I managed a service department some years back where I organized the delivery of parts and heavy equipment, and sent technicians around the globe to service our clients. I prepared the entire job profile and handed the information off to the shipping and receiving departments. The job profile explained the place of origin, timelines, contact information, and a complete list of what was being shipped, including weights. Shipping's role was to gather everything, order the container, contact the shipper, and deliver the product to the shipper. Sounds simple, right? In this case, and for whatever reason, the shipper did not contact the exporter for two days. This delay set everything two days behind schedule. This hold-up caused the container to miss the ship, and the product didn't arrive at its destination for another three weeks. The clients were livid because this set back the entire service scheduling system for the next six months.

This shipper needed to understand the costs and repercussions associated with his decision to not set the schedule in motion in a timely fashion. He also needed to be in sync with all the departments to get that product to its destination on time.

In this case, we had two options: fly the equipment in or find a different shipping method (by ship). Both options added significant routing costs to this order and caused a huge profit loss for our company. This blunder caused a massive amount of frustration on the entire delivery, shipping and receiving team, and added a non-recoverable $5,000 to the service trip.

Every company needs to have a process for delivering, shipping, and receiving parts coming in and product going out, on time, every time. When the people involved are not provided a detailed and accurate order information sheet, or they do not read it, they are not in sync with the process flow, and the profit leaks add up fast.

BE OPEN AND WILLING TO CHANGE

Running a business is hard work! Keeping the sales funnel full, finding the best person for each job, keeping your costs under control, deadlines, stress, the responsibility of leadership, and being able to maximize profits year after year is tough but essential and rewarding in many ways.

The collective theme presented in this book is CHANGE. This means taking a hard look at your business from an objective viewpoint to determine where you're suffering profit leaks, and being willing to make the necessary changes to get back on track. Under the umbrella of CHANGE, there are four areas of focus: Culture, Communication, Structures, and Accountability. How they interlink and their effect on the health of your entire business or organization depends on how the business is managed. Profit leaks come in many forms, but typically, they involve people, time, and money. The secret to minimizing profit leaks is to examine what drives your business toward success—what's truly working—and to determine what areas are holding you back or causing you not to be as profitable and successful as you know you can be. We tend to ignore these areas because it's uncomfortable putting our business (and ourselves) under the microscope, but it's a necessary part of building a strong business with an upward trajectory of profit.

Change is a single act that opens minds, creates opportunity, and ensures that transformation evolves and profits remain constant.

Change involves asking yourself a few hard questions:

> *Is each department or person in your business functioning at top efficiency?*

As you seek to answer this question, note which areas seem to be stumbling.

> *Is a certain person or department being favored or allowed special freedom(s)?*

If so, you are silently sabotaging your progress and profits.

> *Are your departments organized, and do they communicate effectively within and among one another?*

The first answer that comes to mind for most business owners is *yes, everything is going well.* When we look past the comfortable answers and delve deeper into actualities, I am willing to bet that all is not as rosy as it seems.

> *Is it time to remodel your workforce culture so it reflects more of a can-do attitude?*

The cultural makeup of an organization is key to creating progressive attitudes. Make it a priority to create and nurture a culture that is openly collaborative and honest, an environment where respectful conversations are the norm, hard questions can be asked, and responsive answers given. At the same time, strive to diminish negativity during discussions, and be strong when insisting that passing judgment on others is unacceptable. Make sure your employees really understand what is expected of them. Help them understand the importance of respecting their coworkers' skills and abilities, and to hold themselves accountable first, then others.

Once you've guided your staff on the pathway to change, it might be time to upgrade company standards and processes. As the manager or business owner, remember that you manage processes and lead people. Share with your team that you value their work, then introduce the structure you will use to build your systems and processes. As workers develop their skills, grow in self-confidence, and accept their coworkers' abilities, this will encourage each person down the line to raise the bar on their own level of expertise until the service or product is complete. Then, the entire team takes credit. They celebrate each other's accomplishments and the team's success. Profit leaks disappear, and the company flourishes.

As your company reshapes and grows, will you allow yourself to be forced to make changes out of rushed desperation because you've put off those changes for too long?

That's not an optimal situation, and it's also a huge profit leak. Why wait until it is too late? There is no better time than right now to address the problems and get your profit process on an upward trajectory. With a positive mindset, you can move hills. When you combine a positive mindset with a progressive culture, you can move mountains.

CHANGE is essential for organizational growth and success. This shift is the first and last requirement to fixing profit leaks.

STAY CONNECTED WITH THE AUTHOR

Profit leaks can be fatal to running a business. The goal of this book is to offer a new, better, or different understanding that will help you with the day-to-day running of your business. I hope this book has shown the importance of empowerment and the value of embracing change without fear or resistance so that you and your business will grow from strength to strength.

Thank you for taking the time to read my book, *The Profit Process: Building a Successful Business without Profit Leaks*. I would greatly appreciate a positive review on Amazon.com so that other business owners can be inspired and encouraged as well.

Feel free to email me at **brent@pro-bmanagement.com**_or follow me on social media for updates and additional information to build your business:

https://www.facebook.com/probmanagement/
https://www.linkedin.com/in/probmanagement/
https://pro-bmanagement.com/

As a special gift for reading the book, I am offering a 50 percent discount on any services or materials offered by Pro-B Management Inc. When you visit my website linked above, click the Contact Us button and put PROB50 in the subject line of your email. And now, I encourage you to go forth and prosper. You can do it!

Brent Trenholm, CEO
Pro-B Management Inc.

Made in the USA
Monee, IL
08 November 2026

67383844R00069